TABLE OF CONTENTS

SO-AXA-813

FOREWORD

Welcome to the exciting world of VHF amateur radio! The term "VHF" encompasses the range of 100 to 300 MHz, but in this book it has been expanded to include the six meter (50 MHz), 70 cm (420 MHz), 33 cm (902 MHz) and 23 cm (1250 MHz) bands.

If you have a scanner, you'll find a lot of interesting signals in this huge span of frequencies. In addition to TV and FM broadcasting channels, there are police, marine, aircraft, telephone, industry, military and—of course—the radio amateur bands. VHF amateur radio is what this handbook is all about.

VHF radio has a long and interesting history dating back to the days of Hertz and Marconi. And radio amateurs have played their part in exploring and expanding the use of this important natural resource. You, too, can contribute, if you wish.

Written in nontechnical language, this book provides valuable information covering important aspects of VHF radio and tells where you can find additional data beyond the scope of this work.

This handbook is a companion volume to "All About HF (High Frequency) Amateur Radio". Together, they provide comprehensive coverage of amateur communication over the range of 160 meters (1.8 MHz) to 23 cm (1250 MHz). Subjects covered include antennas, propagation, equipment, DX operation, causes and cures of stereo and television interference, and other important aspects of amateur radio often learned "the hard way" through time-consuming and expensive experience. In short, this book tells you how to get your station on the air and how to keep it there!

Note to readers: Band assignments and frequency channels change from time to time depending upon usage and Federal Regulations. Check your local radio club or monitor the channel you propose to use if you are unfamiliar with the local band plan.

Chapter 1

VHF PROPAGATION

Over the Hills and Far Away

Some of the early experiments of Hertz and Marconi took place in the VHF region of the RF spectrum (Fig. 1). However, the success of Marconi's trans-Atlantic communication via long waves in 1901 turned experimenters' thoughts away from the very high frequencies (Fig. 2). Low frequency radio waves were easily generated and detected, while the same could not be said for the "ultra-short" waves. It was not until about 1935 that renewed interest in short range point-to-point communication opened up the VHF spectrum for general amateur and commercial operation. Before then, frequencies above 30 MHz were thought to be of little use. Equipment for those frequencies was hard to build, while results were limited to line-of-sight distances. The waves seemingly were not reflected from the ionosphere.

By 1936, however, amateur interest in VHF radio was growing, and the old 5 and 2-1/2 meter bands were populated by experimenters and rag-chewers, particularly in the metropolitan areas (Fig. 3). It also became apparent that VHF waves were not always limited by the horizon.

After World War II, world-wide frequency assignments were revised and the modern amateur VHF bands came into being. Vast quantities of war surplus VHF equipment spurred interest in the new bands and amateur activity expanded rapidly, at least up to 450 MHz.

A second revolution in VHF amateur radio took place about 1965 when the number of FM channels used for fixed and mobile commercial service

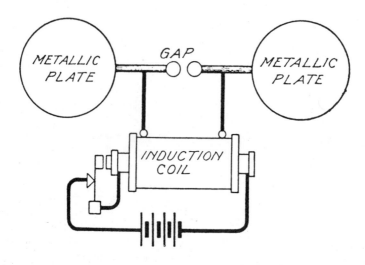

Fig.1 *World's first radio transmitter and receiver were built by Hertz in 1884. The spark transmitter (left) operated near 100 MHz, in the vicinity of today's FM band. A capacitor made of large metallic plates was excited by an induction coil. Radiator was a rudimentary dipole antenna, broken at the center for the spark gap. The receiver (right) consisted of a resonant wire loop with a gap. A tiny spark could be seen in a dark room when the loop was adjusted.*

ALL ABOUT
VHF
(VERY HIGH FREQUENCY)
AMATEUR RADIO

William Orr, W6SAI

RADIO PUBLICATIONS INC.
925 SHERWOOD DRIVE, BOX 247
LAKE BLUFF, IL 60044 U.S.A.

Fig.2 *"The Father of Radio", Guglielmo Marconi worked with centimeter waves, sending his 500 MHz signals between his yacht and land over a distance of 169 miles. His early VHF tests were overtaken by the success of long wave trans-Atlantic circuits established earlier.*

was increased and the channel spacing reduced from 15 kHz to 5 kHz. As a result, a lot of good equipment gradually became obsolete and, because of its low cost, found its way into amateur FM service on the 2 meter band.

In a few years, narrow-band FM became the predominant mode of communication on 2 meters. And 220 and 450 MHz were not far behind. The VHF portion of the radio spectrum was wide open for amateur radio (Fig. 4)!

FM Repeaters Extend the Local Range

In general, low power VHF communication is limited to line-of-sight operation ("what you see is what you get!"). In addition, radio "shadows" exist around buildings and other structures that block VHF waves. Early VHF mobile operation was spotty, with signals swinging up and down as

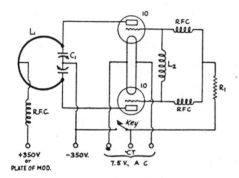

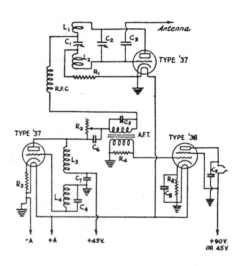

Fig.3 *Push-pull 210 oscillator provided 20-watt signal for serious 1931 VHF operator. When modulated, the simple transmitter sounded good on a "rush-box" receiver. The super-regenerative receiver (right) popularized the 5 meter band. Easy to get working, the receiver used batteries for portable operation. Annoying feature of receiver was mushy, radiated signal which interfered with other nearby "rush-boxes".*

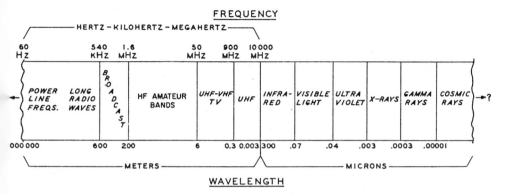

Fig.4 *The electromagnetic spectrum chart is arranged by wavelength and frequency. It shows relationship between electromagnetic waves. At the left (low end of spectrum) are very long radio waves used for long distance, ground wave communication. The regular broadcast band encompasses the medium waves and still shorter waves are useful for ionospheric-reflected, long distance communication. The VHF portion of the spectrum covers 30 to 300 MHz, and the UHF portion covers 300 MHz to 3000 MHz. Gradually, the very short radio waves blend into infra-red waves, and at still shorter wavelengths the waves are visible as light. Shorter than light waves are ultra-violet, X-rays, gamma , and cosmic waves. Scientists suspect that super-long and super-short waves exist in the universe, and the puzzle is to find them.*

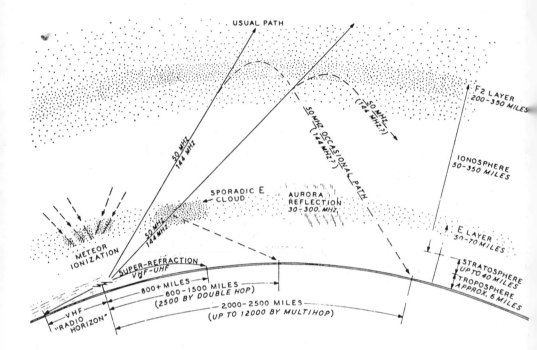

Fig.5 *The atmosphere of the earth is concentrated in a thin layer about 300 miles thick. Ionized layers within this span have the ability to reflect high frequency radio waves. The atmosphere is divided into strata named the troposphere, the stratosphere and the ionosphere. It is in these three layers that interesting modes of VHF propagation take place.*

the vehicle moved along, passing in and out of "shadow" areas.

It was not until the "FM boom" of the mid-1960's that the FM repeater (long used by the commercial services) was adapted and put to use in the amateur bands.

In its simplest form, a repeater is a medium power, automatic FM relay station, usually placed at an elevated location, which expands the limited range of home or mobile stations. FM and repeaters go together like ham and eggs and provide improved, reliable VHF communication which is virtually interference and fade-free. Repeater operation is covered later in this handbook.

Elements of VHF Propagation

While repeaters provide beyond-horizon operation, there are many modes of propagation that provide long distance VHF communication by way of reflected, refracted or scattered radio energy. These modes are of great interest to the VHF operator and are summarized in this chapter.

The Propagation Medium

All terrestrial radio waves are propagated through the atmosphere of the earth, which is considered to reach a height of about 300 miles, as shown in Fig. 5. The atmosphere is not necessary for propagation to take place, as radio waves travel easily through the vacuum of outer space. The atmosphere, however, does exert a profound effect on the radio wave passing through it. The atmosphere is composed mainly of oxygen, hydrogen and nitrogen, but there are traces of many other gasses present, plus a host of foreign materal such as dust, pollen, water, bacteria and fragments from outer space.

The composition of the atmosphere is relatively constant from sea level to the upper reaches, but its density slowly decreases with increasing altitude. At 50 miles, the atmospheric density is much lower than the best vacuum produced on earth.

This thin blanket of air supports all life on earth, protects the earth from deadly radiation from the sun, and provides a medium for long distance radio communication.

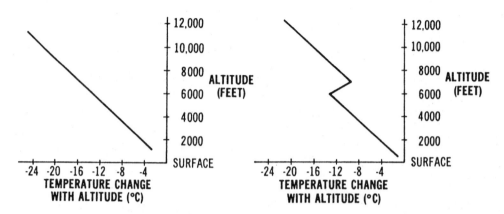

(A) Normal atmospheric lapse rate.   (B) Example of temperature inversion.

Fig.6   Long distance tropospheric propagation takes place because of a temperature inversion. Normal temperature and water vapor content of air decrease with altitude (A). The refractive index of the atmosphere can produce inversion region (B), resulting in an abrupt break in water vapor content. If the inversion is pronounced, the resulting bending of the radio wave will follow the curvature of the earth. Atmospheric ducts have propagated VHF signals over thousands of miles.

Tropospheric Propagation

Tropospheric propagation is the most common form of beyond-the-horizon VHF transmission and is the result of variations in the region of air surrounding the earth.

The portion of the earth's atmosphere extending from sea level to a height of about six miles is named the **troposphere** or "weather layer". It is the home of the winds, storms and rains that continually alter and erode the surface of the earth. Meterological changes in the troposphere are responsible for many interesting VHF conditions.

A radio signal leaving the antenna travels in straight lines. Because of the curvature of the earth, it is soon lost in outer space unless it is bent in some way around the earth. The portion of the wave transmitted along the surface of the earth (the **ground wave**) travels only to the **radio horizon** (about 1.3 times the distance to the optical horizon) before it passes into space.

Point-to-point propagation is a function of the heights of the transmitting and receiving antennas, the distance between them, the terrain, and the power level used. Under normal atmospheric conditions, the signal level over a given path decreases in a predictable manner with increasing distance. Generally speaking, if power, antenna gain, receiver noise and other factors are constant, the range decreases as the operating frequency is increased.

Using low power equipment and a moderate-sized antenna 25 to 30 feet above average terrain and in the clear, it is possible to work similarly equipped stations out to the radio horizon with good reliability on any of the popular VHF bands. Operating with a 10-watt transceiver and a ground plane antenna may be fun, but it does not begin to show what the VHF band is good for! This is why many ardent VHF operators have large antennas, low-noise receivers, big transmitters and try to live in as high a location as possible.

On the other hand, if you wish only to rag-chew with local stations, whatever your radio horizon may be, a well-located repeater can double or triple this range. And this is all some operators are interested in! Thus, normal tropospheric propagation may be considered the main communication mode on the VHF bands.

Tropospheric Bending Propagation

Changes in the normal temperature and water vapor content of the lower atmosphere (troposphere) affect its refractive index and alters the range of VHF waves. For example, after a hot day, a cool evening breeze blowing in from a nearby body of water forces warm air upwards. Or a change in weather patterns brings in a large mass of warm, tropical air which rides over surface air. The result is a **temperature inversion**, with the atmospheric temperature increasing abruptly with increase in altitude (Fig. 6). Such a boundary may extend for 1000 miles or more along a slowly moving weather front. When conditions are right, the path of a VHF wave travelling through the inversion will be refracted enough to match the curvature of the earth. A small inversion near the ground can increase the VHF range up to 400 miles or more. But if the inversion occurs at a height of several thousand feet and extends over a wide area, the increase in range may be over a thousand miles. When the inversion layer is high above the ground

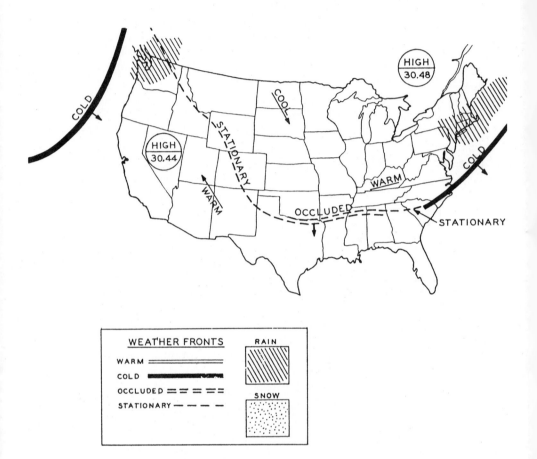

Fig.7 *Temperature inversion forms along a slow-moving weather front where cool high pressure system meets warmer air. A sharp inversion can form along a front at the junction of the air masses, extending as far as 800 miles. Inversions frequently occur along coastal areas bordering large bodies of water and result from onshore movement of cool, humid air.*

and is bound, top and bottom, by the normal atmosphere, an **atmospheric duct** is formed. In some instances these ducts lie between 3000 and 6000 feet above the ground. The ducts have propagated VHF signals over distances in excess of 2500 miles.

At the onset of the duct, the optimum working frequency may be quite high (1500 MHz or so), gradually decreasing as the duct stabilizes and enlarges. If you are located above or below the duct, you cannot take advantage of it. In addition, coastal over-water ducts seldom extend very far inland.

The most common VHF ducts exist over the California-Hawaii path, off the coast of Australia, over the Gulf of Mexico, the Mediterranean Sea, the Bermuda-US path and around the Canary Islands.

Temperature Inversion

Temperature inversion effects increase with frequency. A few effects have been observed as low as 50 MHz, but ducts are much more frequent from 144 MHz to at least 10,000 MHz (10 gHz). In some areas of the world (the Indian Ocean, for example), ducts are so frequent that they constitute a "normal" condition.

Temperature inversions occur almost every evening along the Atlantic, Pacific, and Gulf coasts of the USA (and along the ocean coasts of other countries). Coastal VHF stations often work each other over distances up to 1000 miles, with occasional contacts over distances as great as 2500 miles. For non-coastal stations, the most spectacular tropospheric communications occur during the fall months when warm, hazy, calm air days are followed by cool nights. Slow, easterly-moving masses of cool Pacific air meet warm Gulf air. The two air masses may preserve their separate identities for days as they move eastward across the continent, before they combine into a homogeneous mass (Fig. 7). Signals are strong in the late evening and again in the early morning around sunrise. Long distance VHF communication is possible under such conditions.

Besides enhanced VHF propagation, fog, smog and high haze are also indications of temperature inversions. Heavy, warm air that traps air pollutants and holds them close to the ground seems to do the same to VHF signals.

Good indicators for an inversion are the appearance of VHF and UHF television signals from outside the normal reception area. Most often the signal is quiter strong, depending on the station's location with respect to the duct.

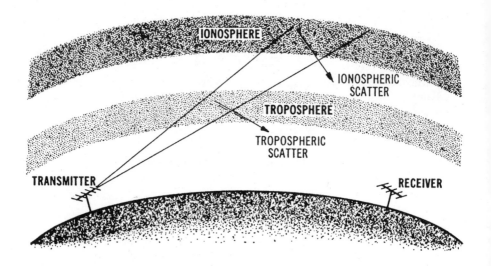

Fig.8   *VHF forward scatter provides beyond-horizon radio range. Tropospheric turbulence provides scatter path up to about 500 miles and ionospheric scatter provides path ranging from 600 to nearly 1400 miles. The outstanding characteristics of scatter communication is that the average signal level is very constant and predictable, regardless of weather or ionospheric conditions.*

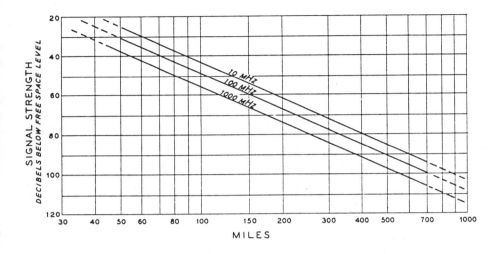

Fig.9 *Tropospheric scatter communication is possible up to about 500 miles. Path attenuation increases gradually with frequency but rapidly with respect to path length. There is no theoretical limit to signal range.*

Tropospheric Scatter Communication

The advent of high transmitter power, large beam antenna systems, and more sensitive receiving equipment revealed that a VHF signal can travel for a suprising distance beyond the horizon. The outstanding characteristic of such a signal is that the average strength is very weak but quite constant regardless of weather or ionospheric conditions. This is called **tropo-scatter** propagation because the radio wave is throught to be reflected from random irregularities which are always present in the lower atmosphere. These irregularities change the index of refraction sufficiently to cause faint signal illumination well beyond the horizon. The phenomenon is similar to the way the overhead beam of a searchlight can be seen as a glow beyond the horizon (Fig. 8).

There is no theoretical limit to the range of scatter propagation, but practical considerations seem to limit amateur scatter communication to about 500 to 700 miles. The path attenuation increases gradually with

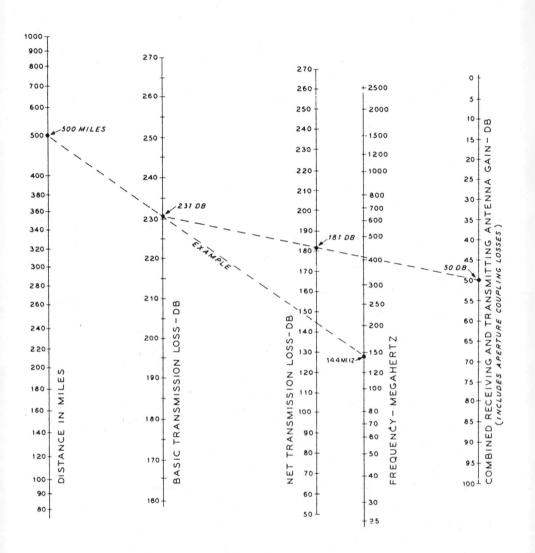

Fig.10 *VHF nomograph is used to determine circuit re-quirements for tropo scatter path. As shown, a 144 MHz, 500 mile path has a transmission loss of 231 dB. If the combined receiving and transmitting antenna gain is 50 dB, the net transmission path loss drops to 181 dB. The attenuation figure will vary with the topography of the path, but this chart provides good accuracy even over hilly terrain.*

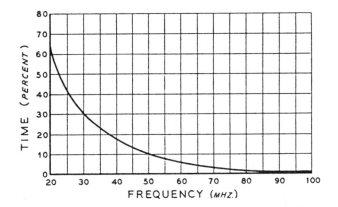

Fig.11 *The percentage of time that sporadic-E propagation occurs during the summer months is a function of the frequency.*

frequency, but rapidly with respect to path length (Fig. 9). Circuit requirements for a representative tropo path are shown in Fig. 10.

Scatter signals are weak at best and experience has shown that CW is the best mode for communication via this propagation mode. Antenna location and gain, high power, and a low-noise (sensitive) receiver make tropo-scatter an interesting world of weak-signal DX.

Sporadic-E Propagation

Sporadic-E propagation (abbreviated Es) is caused by clouds of ionized atoms in the E-layer of the ionosphere, that region about 50 miles above the earth which reacts to radiation from the sun. The clouds are generated in the lower latitudes and generally drift in a northwesterly direction.

The creation of ionized clouds is not understood but is thought to be related to wind shear and lightning storms. While normal E-layer reflection of HF waves is quite predictable, sporadic-E VHF propagation is highly unpredictable and is responsible for the greatest percentage of 6 meter long distance communication (Fig. 11). The DX range on this band supported by Es propagation ranges from about 800 to 1200 miles. Double-hop Es may extend the range out to 2500 miles. Signals are usually quite strong,

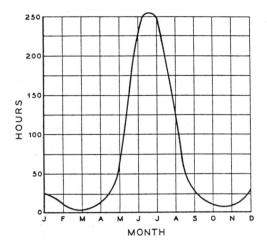

Fig.12   *Sporadic-E openings in the 50 MHz band tend to peak during the summer months. Openings up to 1400 miles are common throughout the U.S. and from the southern states to Mexico and Central America. An increase in sporadic-E openings are observed during periods of low sunspot activity, decreasing during the higher portion of the sunspot cycle.*

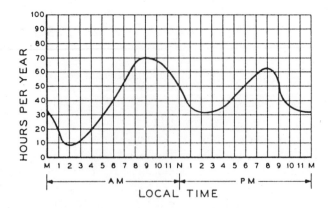

Fig.13   *Sporadic-E openings tend to take place during the hours between 8 to 11 a.m. and 6 to 8 p.m., local time.*

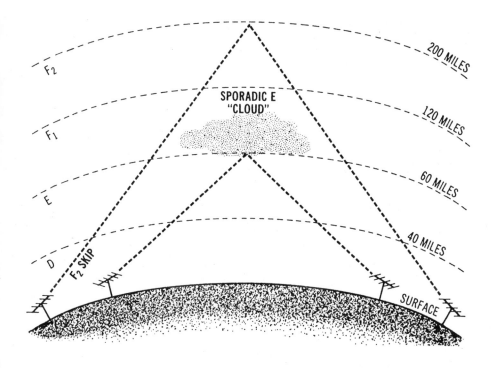

Fig.14 The E and F2 layers are of interest to the VHF operators, since this is where reflection at VHF occurs. The E-layer occasionally becomes charged, and patches of "ionic clouds" reflect both HF and VHF waves. The F2 layer can reach a sufficient ionization level to reflect 50 MHz signals when the sunspot cycle is high. The 1958 sunspot peak provided periods of extended 50 MHz DX conditions, which were observed to a lesser extent during the 1969 and 1980 peaks.

but may vary rapidly in strength from minute to minute. Openings tend to be most numerous during the summer months (Fig. 12).

Sporadic-E openings are occasionally observed on the 2 meter band, usually during July and August. Some sporadic-E propagation has been noted in the 220 MHz band, but it seems to be quite rare at this frequency and above.

A good alert for VHF Es skip is the presence of short-skip conditions on the 10 meter band. When stations as close as 300 to 500 miles can be heard on 10 meters, the 6 meter band may be open for Es propagation. And when the 6 meter Es skip is as short as 500 miles, the 2 meter band may be open for Es long distance contacts. Openings tend to take place in the early morning and evening hours (Fig. 13).

Experienced VHF operators also monitor TV channels 2 and 3. When picture interference is noted from another station, it is a clue that Es skip is present.

VHF Ionospheric "Skip" Propagation

Ionospheric skip (F2 layer propagation) supports long distance communication in the HF region. Occasionally, during the years of high sunspot numbers, the upper atmosphere is sufficiently ionized to support DX "skip" communication up to 50 MHz (Fig. 14). Typically, the sunspot count must be higher than 125 (corresponding to a solar flux of 175) to allow such propagation. Since the sunspot count is not predicted to reach this figure for the rest of the 20th century, it is probable that this exciting mode of VHF long distance communication will be rare in the immediate future! (The effects of the sun and sunspots on radio communication are discussed in detail in the companion handbook, "All About HF (High Frequency) Amateur Radio", available from Radio Publications, Inc., Box 247, Lake Bluff, IL 60044).

Ionospheric Scatter Propagation

Ionospheric scatter has often taken place on the 50 MHz band and has been noticed on the 144 MHz band. The scatter mode occurs in the D-layer of the ionosphere and the signal range seems to be 500 to 1200 miles. Signals are most reliable on an east-west path (Fig. 15).

Because only a small part of the scattered energy returns to earth, the signals are very weak. Large antennas, high power transmitters and low-noise receivers are required for consistent CW communication via this mode.

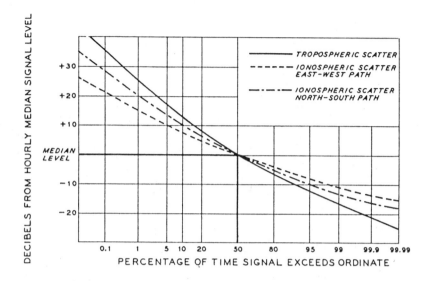

Fig.15 *Tropospheric and ionospheric scatter propagation prove most reliable on an east-west path. Rapid signal fading is caused by multipath transmission, with the rate of fade increasing as either frequency or path length is increased. Slow fade over hours or days is a result of atmospheric variations along the path.*

Aurora Reflection Propagation

Aurora displays are almost constantly visible during the evening hours around the magnetic poles (Fig. 16). The strongest displays are seen around the equinoxes in March and September and begin in the early evening hours. Extremely strong auroral displays may be seen as far south as Georgia, but because the magnetic north pole is located near Thule, Greenland, western areas of the United States and Canada see fewer aurora displays than do corresponding eastern areas (Fig. 17).

Aurora reflection takes place between 50 to 80 miles above the earth and is observed in the 6 and 2 meter bands, and occasionally in the 220 MHz band. The reflected path length may be as long as 1200 miles. The aurora reflected signal has a peculiar "buzz" to it, which makes SSB communication difficult, and most amateurs use CW during periods of aurora activity (Fig. 18).

Fig.16   *Aurora covers Canada and portions of the United States in composite photo taken from U.S. weather satellite. Lights of cities stand out in this night view. Aurora band is about 600 miles across and pulses in response to energy received from the sun. Color most often seen from earth is bluish-green. Colors result when oxygen and nitrogen atoms in atmosphere are bombarded by incoming protons and electrons from the sun.*

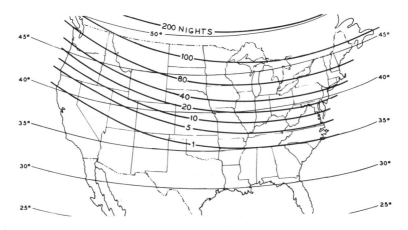

Fig.17 *Aurora display is most prevalent at northern latitudes, but on occasion can be seen as far south as New Mexico or Arkansas. Aurora reflects VHF signals which are modulated by rapid oscillation of aurora, resulting in a characteristic "growl" or "hiss" superimposed on the signal.*

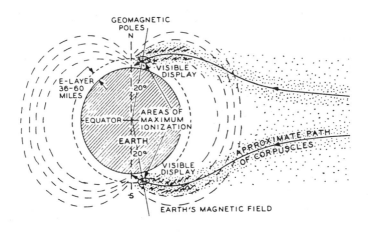

Fig.18 *Maximum aurora display is concentrated in a broad belt near 70 degrees latitude from each geomagnetic pole. The visible ionization occurs at a height of 60 to 70 miles in the E-layer of the ionosphere.*

Fig.19   *Aurora display may signify east-west transmission paths are open up to 500 miles on 50 and 144 MHz bands. The aurora is caused by emission of charged particles from the sun.*

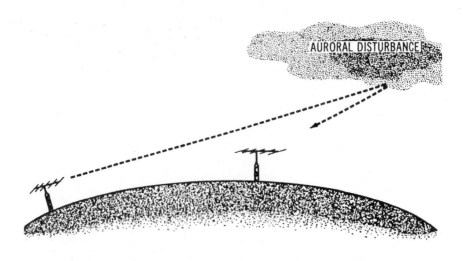

Fig.20   *VHF operators aim their antennas north during an aurora opening. Once the opening is underway, it is possible to aim east or west of north for maximum range.*

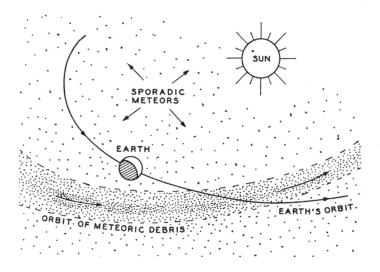

Fig. 21 Earth cuts through a stream of meteor debris in path of a comet. Width of the debris orbit may require several days for the earth to move through the path.

An indication of aurora conditions is the presence of wavering or "watery" sounding signals on the HF bands. In addition, aurora propagation can be detected on a TV receiver on some channels by listening for a "buzz" on the audio signal (Fig. 19). Once the opening is underway, it is often best to aim a little east or west of North for maximum east-west operating range (Fig. 20).

Meteor Scatter Propagation

While the earth is under a continual shower of meteors from outer space, large meteor showers occur at various intervals during the year (Fig. 21). As the meteors flash through the sky, their trails produce a "tunnel" of ionization which is capable of reflecting radio signals (Fig. 22). In general, the number of trails tend to increase slowly during the evening hours, reaching a peak just before sunrise, decreasing as the morning advances. The peak months for **meteor scatter**propagation in north America are June, July and August.

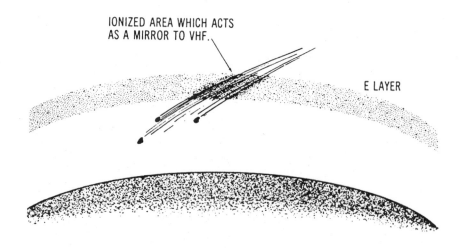

Fig.22   Meteor debris result in ionization of atmosphere. Large cloud of meteors yield a burst of signals lasting for several seconds. Bursts as long as several minutes have been noticed when several large meteors fall together. Over 12 billion meteors are swept up by the earth each 24 hours, most of them microscopic in size.

Heavy meteor showers of short duration occur approximately August 10-14 (Perseids) and December 10-14 (Geminids), as shown in Table 1. These provide good DX opportunities on the 6 and 2 meter bands. The duration of a meteor trail is short, perhaps 10 to 30 seconds, and often a scatter signal appears only as a few words of SSB or a burst of CW from a station up to 1200 miles away. A high speed exchange of signal reports on prearranged schedules is commonly used by amateurs to take advantage of this erratic form of propagation.

Transequatorial Scatter Propagation

Transequatorial (TE) scatter was first noticed by radio amateurs in 1947. In the spring and fall months, amateurs located in broad belts, roughly 1500 to 2500 miles on either side of the geomagnetic equator, are able to communicate with each other via this mode (Fig. 23). Propagation is roughly north-south in direction and takes place during the late afternoon or early

Shower Name	Average Peak Date	Usable Dates	Radio Bursts Per Hr	Optimum Time*	Path	Notes
Quadrantids	Jan 3	Jan 1–4	50	0200–0800	NW-SE	Intense shower Fast-moving meteors
				0800–0900	S	
				0900–1430	SE	
Aquarids	May 4	May 1–6	15	0500–0600	NE-SW	High-speed particles
				0630–0830	E-W	
				0830–1000	NW-SE	
Arietids	June 7–8	June 1–15	60	0500–0700	N-S	Intense daytime shower Small slow-speed particles
				0715–0900	NE-SW	
				0900–1000	E-W	
				1030–1200	NW-SE	
				1230–1415	N-S	
Perseids	Aug 12–13	Aug 10–14	50	2230–0400	NW-SE	Many large meteors One of the best showers
				0500–0700	E-W	
				0800–1300	NE-SW	
Orionids	Oct 21–22	Oct 18–23	20	0030–0130	N-S	
				0130–0330	NE-SW	
				0330–0500	E-W	
				0500–0730	NW-SE	
				0730–0830	N-S	
Leonids	Nov 16–17	Nov 14–18	15	0200–0330	N-S	
				0330–0530	NE-SW	
				0530–0700	E-W	
				0700–0900	NW-SE	
				0900–1100	N-S	
Geminids	Dec 12–13	Dec 10–14	60	2100–2330	N-S	Very reliable One of the best showers
				2330–0130	NE-SW	
				0130–0230	E-W	
				0230–0400	NW-SE	
				0400–0730	N-S	
Ursids	Dec 22	Dec 21–22	15	2300–1600	S	

* Optimum time is local time at midpath.

Table 1. Meteor showers of interest to VHF operators are listed above. The spring showers peak between midnight and 6 a.m., and again about noon. Meteor activity tends to build up and drop off slowly so operators usually conduct tests a few days either side of the peak.

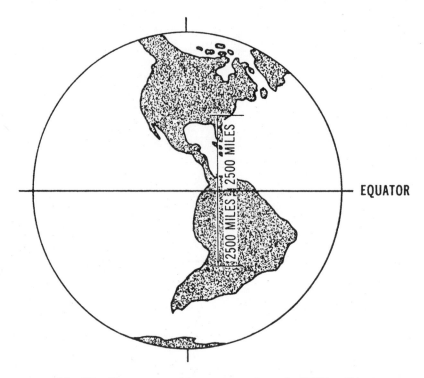

EQUATOR

Fig.23   Six meter stations approximately 2500 miles on either side of the equator may be able to communicate via transequatorial scatter. Propagation occurs during periods of high solar activity in late afternoon and early evening hours. It is most frequent during spring and fall months.

evening hours on the 50 MHz band. This mode is generally associated with high sunspot numbers.

In the western hemisphere, the geomagnetic equator runs about 15 degrees south of the true equator, consequently only stations in the southern area of the U.S. are optimally located to take advantage of TE propagation. Stations in South Africa and the Mediterranean, as well as in Japan and Australia, can make favorable use of this unusual mode of propagation.

TE scatter results in strong signals having a rapid flutter. Small antennas and medium power are effective when this transmission mode opens up the 6 meter band for DX.

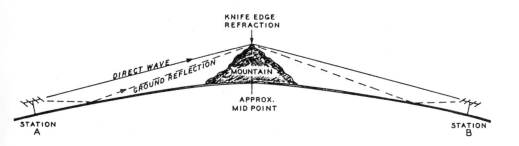

Fig. 24 Knife-edge refraction by midpoint obstacle provides signal boost on long VHF path. Obstacle height of about 5000 feet boosts VHF signals at a distance of about 100 miles from point of refraction. The received signal is many decibels stronger than if obstacle were removed, leaving only normal earth curvature.

Knife-edge Diffraction

Under certain conditions it is possible for a ridge of hills or mountains to refract VHF waves over their crests (Fig. 24). To aid VHF propagation, the crest of the hill should ideally run at right angles to the signal path, have a sharp break, and be situated about halfway between the stations attempting to communicate. A ridge height of about 5000 feet will boost VHF signals at a distance of about 100 miles from the point of refraction.

An apparent barrier to VHF communication between two points can often be overcome if a high hill off the direct path between the locations is visible at both locations. Operators can point their antennas at the hill and use it to refract their signals around the apparent obstruction (Fig. 25).

Tropospheric Backscatter Propagation

Tropospheric **backscatter** propagation was discovered when return echo signals from distant ground were first observed on radar. This effect can support communication between stations at critical distances apart which cannot be covered by other propagation modes. The effect is common on both the 6 and 2 meter amateur bands but it has been noted up to 1250 MHz. The stations involved aim their antennas at a common point on the

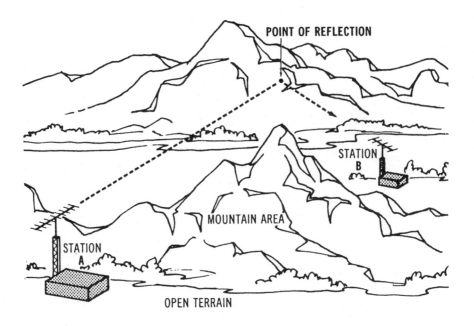

Fig.25 *Operators in hilly areas can often reflect their signals to a distant station by aiming at a hill or peak that is not in the direct path. A sharp ridge refracts the signals, bending them back over an area where the signals are normally blocked from line-of-sight propagation.*

horizon and copy each other's backscatter signals. The distances covered by this communication technique are usually a few hundred miles, but well-equipped 144 MHz stations over 800 miles apart have communicated via backscatter with weak but stable signals. CW is the best form of communication.

Moonbounce Propagation

Communicate via the moon? Shortly after World War II amateurs in the military heard echo signals from the moon on a specially modified radar transmitter/receiver. Eventually amateurs using home stations achieved two-way contact by reflecting their signals from the moon. This interesting mode of communication is covered in a later chapter.

Chapter 2

The VHF BANDS— AN OVERVIEW

A Band-by-Band Look at the VHF Spectrum

Each VHF amateur band has its own unique characteristics. This chapter discusses the individuality of each band, and the pecularities that make that band interesting and useful. Certain bands lend themselves more readily to some modes of propagation than do others (Table 1). In addition, as operating conditions and band occupancy change, mode allocations for a particular band are therefore also subject to change. This provides a rich field for experimentation.

6 Meters (50.0 to 54.0 MHz)

This interesting band can demonstrate both HF and VHF propagation characterisitcs. Tropo bending, sporadic-E, aurora, and F2-layer reflection propagation modes all occur, sometimes several modes at a time (Fig. 1). In low latitudes, transequatorial propagation adds excitement for the 6 meter operator. Sporadic-E propagation provides intercontinental DX with many trans-Atlantic and West Coast to Hawaii contacts logged during good openings. DX opportunities to the Caribbean area also frequently exist.

Regional **beacon stations** are in daily operation throughout the world, providing a means of determining DX propagation over many paths. Beacon frequencies range from 50.025 MHz to 52.5 MHz. Most beacons are CW and use an omindirectional antenna. They identify themselves with the call sign of the country or state in which they are located.

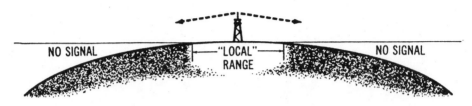

(A) Line of sight.

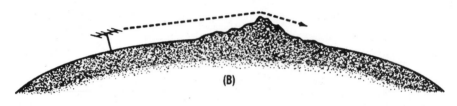

(B) Knife-edge refraction.

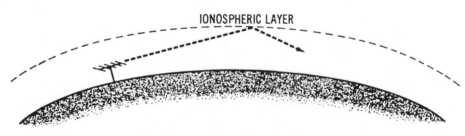

(C) Ionospheric reflection.

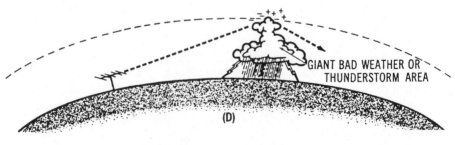

(D) Atmospheric disturbance.

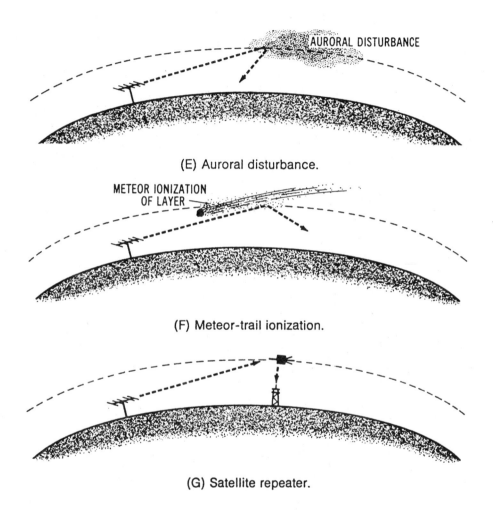

(E) Auroral disturbance.

(F) Meteor-trail ionization.

(G) Satellite repeater.

Fig.1 Propagation modes of the VHF bands. All the bands support interesting ways of sending signals that extend line-of-sight communication. Often several unusual modes occur in a band as the same time.

Propagation Mode	50 MHz Band (6 Meters)	144 MHz Band (2 Meters)	220 MHz Band (1¼ Meters)	420 MHz Band (70 Centimeters)	1215 MHz Band
Ground & Direct Wave.	150 mi Typical	100 mi Typical	80 mi Typical	80 mi Typical	50 mi Typical
Scatter	Good 500 mi	Good 250 mi	Fair	Poor	Very Poor
Meteor Scatter	Excellent 1500 mi	Excellent 1500 mi	Good	Poor	None
Aurora	Excellent	Excellent	Good	Fair	None
Tropo	Poor	Good	Excellent	Excellent	Excellent
F2	Excellent During High Solar Activity	None	None	None	None
Sporadic E	Excellent	Good but Rare	None	None	None
EME	Possible, Difficult	Good Activity	Some Activity	Good Activity	Some Activity
Satellite Allocation		Yes		Yes	

Table 1 A summary of propagation characteristics of the VHF amateur bands. The six meter band provides F2 "skip" propagation during summer and fall of years having a high sunspot count. Sporadic-E is common on this band during the same months and has also been noticed on rare occasions on the 2 meter band. All of the bands provide interesting modes of long distance communication.

Fig.2 "Herringbone weave" pattern on TV screen is caused by interference from nearby FM transmitter. The pattern varies with the modulation on the signal. This interference may appear on TV channel 2 from nearby 6 meter station.

In some areas 6 meter operation is handicapped by a powerful, adjacent channel 2 TV transmitter which fills the high frequency end of the band with unwanted picture modulation. In addition, 6 meter operators often have problems caused by overloading the front-end of a TV receiver tuned to channel 2 (Fig. 2). Thus, the band bears an uneasy relationship with TV in those areas having channel 2 activity.

CW operation takes place below 50.1 MHz and SSB operation occupies the lower portion of the band (50.1 to about 50.3 MHz) with a "DX window" between 50.1 and 50.125 MHz. During the evening hours, SSB contacts up to 100 miles are commonplace with a 10 watt transceiver and a well-located antenna.

FM and associated repeater operations encompass the range of 51 to 54 MHz. The popular **simplex** (point-to-point) calling frequency for local FM operation is 52.525 MHz. Radio control of models generally occurs above 53 MHz.

Communication that makes use of a space satellite is less dependent on the whims of the atmosphere that support conventional terrestrial communication. Satellites for amateur radio have been supplied by various amateur non-profit organizations. The "birds" have been built by hams all over the world and are coordinated by AMSAT (the Radio Amateur Satellite Corp.) Details in Chapter. 5

During long distance band openings the lower portion of 6 meters becomes alive with signals, and pileup conditions can exist when a rare state or country comes on the air. CW operation peaks around 50.05 MHz and SSB about 50.11 MHz. Band openings can last from a few minutes to a few hours. Checking 6 meters for sporadic-E DX openings during periods of short skip on 10 and 15 meters can be very rewarding for the alert operator.

2 Meters (144.0 to 148.0 MHz)

The 2 meter band is the "work horse" of VHF amateur radio. It is thickly populated with FM repeaters while supporting SSB and CW communications. Experimental communication by way of the moon (**moon bounce**) is popular on 2 meters, as well as communication via man-made satellites orbiting the earth. Many amateurs who enjoy weak signal DX'ing run maximum power and make use of very high gain antennas to employ the rarer modes of communication available on this band.

The low end of the band from 144.0 to 144.1 MHz is reserved for CW operation, with SSB operation from 144.1 up to about 144.3 MHz. The SSB calling frequency is 144.2 MHz. **Packet radio** and amateur teletype operations take place at or near 145.01 MHz. FM repeater operation starts at 144.6 MHz (with point-to-point **simplex** operation at selected frequencies), and occupies the remaining portion of the band.

1-1/4 meters (220.0 to 225.0 MHz)

This band is somewhat similar to the 144 MHz band, supporting most of the same modes of communication. Various types of tropospheric propagation occur, however no ionospheric transmission modes have been observed.

The CB/SSB calling frequency is 220.1 MHz and FM operation via repeaters is popular above 222.34 MHz, with CW and SSB operation below this point. Simplex operation occurs between 223.40 and 223.90 MHz, the most popular frequency being 223.50 MHz.

In metropolitan areas the band is as crowded as the 2 meter band. The operating range of low power equipment using an antenna in the clear is roughly the same as on 2 meters. As antenna size is much smaller than an equivalent 2 meter setup, high gain arrays for this band can be constructed inexpensively and occupy little space.

70 cm (420.0 to 450.0 MHz)

The operating range of low power 70 cm equipment with good antennas in the clear is close to the range offered by the 1-1/4-meter band.

Long distance thermal inversion propagation has been noted, with occasional trans-oceanic contacts between California and Hawaii. Contacts across the Gulf of Mexico have been made, but they are less common because of a lack of activity south of the U.S. border. Some "moonbounce" stations operate on the 70 cm band because a high gain antenna occupies relatively little space and high power can be generated from conventional VHF tubes. Amateur communication satellites also are operational on this band from time to time.

The CW/SSB calling frequency is 432.1 MHz. The band is rapidly becoming occupied by FM repeaters above 442.0 MHz, with **simplex** frequencies around 446.0 MHz. Some amateurs experiment with wideband television transmission and other unusual communication systems around 438.0 MHz. One of the widest amateur bands, it is becoming congested in urban areas of the country.

33 cm (902.0 to 928.0 MHz)

The 902 MHz band is a secondary service band for amateurs. That is, interference from other services operating in this band must be accepted by amateurs. Industrial equipment operates near 913 MHz and government services have channels in this region. Additionally, in some states the band is not available at all. Check with your local radio club, the F.C.C., or the A.R.R.L. for the latest information.

Thermal inversion propagation has been noted on these frequencies, otherwise propagation seems similar to that of the 23 cm band.

Amateur SSB transmissions take place below 903.5 MHz and the frequency 903.1 MHz is suggested as the national CW/SSB calling frequency. Repeaters operate above 907.0 MHz.

23 cm (1240 to 1300 MHz)

This band was unused for many years until low noise transistors and phase-lock frequency control permitted the use of channelized FM equipment. Tropospheric bending and ducting provides exciting long distance contacts. Aurora reflection has also been noted on this band. Generation of high power at these frequencies is difficult, and this has restricted use of the band as far as moonbounce or other forms of long distance communication are concerned.

One popular CW/SSB calling frequency is 1296.1 MHz. More and more FM repeaters are being installed on the band above 1270 MHz, and popularity of this communication mode is continuing to grow. The FM **simplex** frequencies range from 1294 to 1295 MHz, with 25 kHz channel spacing. The most popular **simplex** frequency is 1294.5 MHz. Two amateur television (ATV) segments are located in the band, interspersed with FM point-to-point and digital channels.

At these frequencies, one limiting factor with regard to weak signal DX communication is the noise generated in the receiver itself. In addition, precipitation or moisture in the air can cause noticeable signal attenuation even over a relatively short path. Typical range for low power communication with a good antenna is 20 to 40 miles. Propagation is more terrain-sensitive than on the lower bands as trees and other foliage absorb signals to a greater degree in this frequency range.

Note that 'RF generated at these frequencies can be injurious to your eyes, so do not point an antenna in the direction of your or anyone else's eyes, and do not get your face near high power amplifiers.

Operators have observed that a layer of rain or snow on a high gain 1250 MHz antenna can severely detune it. On the other hand, the same moisture effects that cause antenna problems can also produce tropospheric beyond-the-horizon propagation.

Antenna Polarization

With most forms of VHF propagation, it is important that the radio wave polarization be the same at each end of the circuit. This requires that the active elements of the receiving and transmitting antennas lie in the same relative planes with respect to the earth. The two most commonly used planes of polarization are horizontal and vertical (Fig. 3). Only during the relatively rare instances of VHF F2 ionospheric skip on the 6 meter band is there enough depolarization of the wave to make antenna orientation unimportant.

Vertical polarization is used extensively in local, short distance FM communications where universal coverage is required. Small, non-directional antennas are easy to build with vertical elements, especially in the case of mobile installations. A quarter wavelength whip, with the metal vehicle top serving as a ground plane, is a very popular and effective VHF mobile antenna.

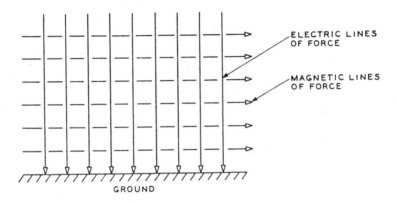

Fig.3 If you could see a radio wave, perhaps this is what you would observe if the wave was travelling out of the page towards you. The arrow heads indicate the direction of the lines of force. Here, the lines of the magnetic field are parallel to the ground and the lines of the electric field are vertical. By definition, this wave is vertically polarized. This drawing represents a wave leaving a vertical antenna. The field surrounds the antenna, so this illustrates just a small portion of the wave front.

Horizontal polarization is preferred for long distance communication. It also provides some discrimination against local, man-made noise, since most electrical noise seems to be vertically polarized. The CW and SSB enthusiasts prefer horizontal polarization as these modes do not enjoy the noise rejection afforded by FM operation, making the noise rejection of horizontal polarization a plus.

Since most VHF antennas are quite small, many amateurs who work all modes have separate antennas for horizontal and vertical polarization, and often switch from one to the other.

Amateurs interested in satellite communication often favor circular polarization because it receives and transmits horizontal and vertical polarization equally well, and the polarization of a satellite signal often changes rapidly. In addition, horizontal or vertical antennas receive circularly polarized signals without difficulty.

Chapter 3

THE VHF REPEATER AND HOW IT WORKS FOR YOU

A VHF FM repeater is a station which automatically retransmits incoming signals. The great popularity of VHF FM, particularly in areas where propagation is often poor (mountainous or hilly country), is due mainly to the use of repeaters which can turn a weak, noisy path into a good one, or a poor location into a favorable one. Repeaters are especially useful for mobile operators, and for operators using hand-held transceivers.

Most repeaters use two frequencies in one amateur band. The first is the **input frequency** which receives incoming signals, and the second is the **output frequency** which retransmits the incoming signal. If the repeater is at a high elevation above the surrounding terrain, has a good set of antennas and medium power level (50 to 100 watts), it may offer coverage of 100 miles or more. Antenna polarization of the repeater is usually vertical to accomodate mobile operators.

Specialized repeaters exist for amateur television, packet radio, teletype and other modes. Since they are few in number, they are not discussed in this handbook.

While operating through a repeater, your station receives or transmits, but not both at the same time. This is termed **half-duplex**. You listen on the output frequency of the repeater and transmit to it on the input frequency. The frequency changeover is accomplished automatically by transceiver circuitry and the **press-to-talk** switch on the microphone.

VHF amateur repeater at a remote mountain-top installation in California. The repeater takes advantage of height to extend operating range of low power, line-of-sight mobile operators. Repeater can turn a weak, noisy signal path into a good one, or a poor location into a favorable one. The repeater is usually unattended and means must be provided to control and monitor the equipment.

Some two meter repeaters have the output frequency lower than the input frequency, while others have the output higher. The user must select the proper frequency shift for his transceiver. Most equipment provides switchable up- or down-transmitter shift, usually designated as **plus or minus**.

An example of **full duplex** is the telephone, where both parties may receive and transmit at the sime time. Regular two-way amateur operation without benefit of a repeater is termed **simplex**, as each operator takes a turn in transmission.

Why FM?

Frequency modulation has certain characteristics (as compared to SSB) that make its use with the VHF repeater very effective.

FM provides a noise-free quality which is insensitive to most man-made noise and atmospheric static. In addition, the strongest signal on the channel **captures** the receiver, and weaker signals are not heard. This provides interference-free service. Finally, the receiver **squelch system** silences the receiver when there is no received signal. All of these advantages have great appeal, especially for the mobile operator.

FM proved its worth in North Africa in World War II when VHF repeaters were set up covering the coast from Tunisia to Algiers. This permitted instant, interference-free communication between the various Allied armies locked in the struggle with Rommel's forces.

A Quick Overview of FM

An FM transmitter provides a carrier which remains at a constant level. When modulation is applied (the voice, for example), the carrier is varied in frequency or phase above and below the center frequency in accordance with the modulation. The bandwidth of an FM signal is a function of the frequency of the modulating signal as well as its amplitude. The FM process generates sidebands (as does SSB), and the **frequency deviation** of the transmitter results in a signal occupying a band of frequencies. For maximum utilization of a channel, the deviation of the transmitter should match the passband "window" which is determined by a filter in the receiver.

The FM receiver is also equipped with a **limiter** stage to eliminate unwanted noise, and to hold the amplitude of the received signal below a maximum level. A **squelch** circuit silences the audio stages when there is no signal.

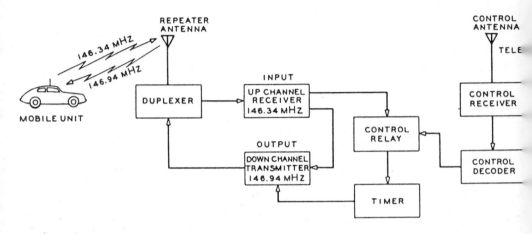

Fig.1 *The VHF FM repeater receives and amplifies signals on one channel and retransmits them on another channel. In this example, the repeater input channel (**up-link**) is 146.34 MHz and the output (**down-link**) is 146.94 MHz (a "34-94" pair).*

*The repeater is connected to the antenna through a **duplexer** which isolates the receiver and transmitter from each other. The repeater system is turned on and off by means of a control relay and timer which respond to a coded signal sent to the unit by telephone line or VHF link. The timer is provided for fail-safe operation. The control system is at the right of the illustration.*

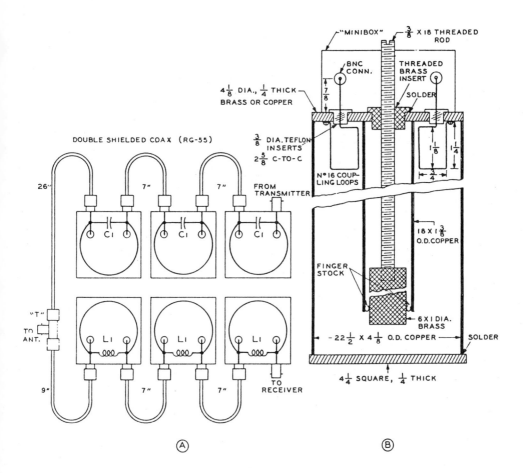

Fig.2 *High performance coaxial filters serve as duplexer for 2 meter repeater. The duplexer provides isolation between receiver and transmitter when both are connected to a single antenna. (A) Three capacity-tuned filters (top) pass transmitter signal and reject receiver input signal. Three inductive-tuned filters (bottom) pass received signal and reject signal from transmitter.*

The filters are adjusted with aid of a signal generator and detector for proper attenuation and passband characteristics. (B) Section of typical filter unit. High-Q, coax cavity is silver plated for best performance. Cavity functions as a resonant circuit.

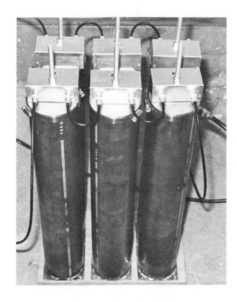

Fig.3 Duplexer built by Michigan City, IN 2-meter repeater group to permit using single antenna for transmission and reception. Duplexer has insertion loss of under 2 dB and an attenuation in excess of 95 dB at the 600 kHz rejection frequency.

The FM transceiver incorporates both transmitter and receiver in one unit. Circuits common to both units are generally combined in a single IC chip to save weight, space and power. A single frequency determining circuit functions for both receive and transmit modes, insuring that the units are locked on the chosen channel.

Repeater Operation

A representation of a repeater is shown in Fig. 1. It is an interconnected receiver and transmitter, with appropriate control equipment. To prevent the transmitter from interfering with the nearby receiver, special devices called **duplexers** are used. These are extremely selective and rather costly filters which isolate the input of the receiver from the output of the transmitter (Fig. 2).

The task of the duplexer is to make sure that no emission from the transmitter reaches the receiver which, on the 2 meter band, is tuned only 600 kHz away. A very high-Q (highly selective) filter can produce a rejec-

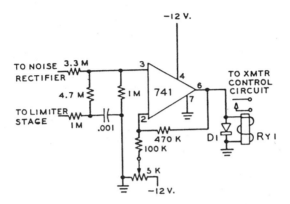

Fig.4 *A carrier-operated relay (COR) activates transmitter portion of the repeater when an input signal of adequate strength is present. Signal-operated relay is held open by rectified noise taken from discriminator circuit of the receiver. When a carrier is received, quieting the receiver, the noise level drops and relay closes. Threshold control is set by compensating voltage derived from limiter stage of receiver.*

tion of over 30 decibels (a power reduction of 1000 times) to the transmitted signal at the input of the receiver. Several such filters are combined into a duplexer (Fig. 3), permitting both receiver and transmitter to simultaneously use the same antenna.

Duplexers are expensive and less costly alternatives exist. One is to use separate antennas for receiving and transmitting and space them apart sufficiently so that a single filter will do the job. Typically, at 146 MHz with a 600 kHz difference in frequencies, a well-shielded repeater may require antennas separated 300 feet in the horizontal plane, or 40 feet in the vertical plane (assuming vertical polarization is used).

When repeater antennas can be separated by about a mile, the overload problem disappears. In such an installation, the receiver and transmitter portions of the repeater are joined by a telephone line or radio link above 220.5 MHz. The repeater itself may be controlled by such a link.

In large cities, a secondary advantage of separate receiver and transmitter repeater sites is the presence of many high power commercial VHF TV and FM transmitters and other signal sources that saturate the area

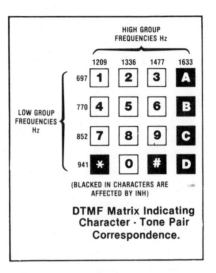

DTMF Matrix Indicating Character - Tone Pair Correspondence.

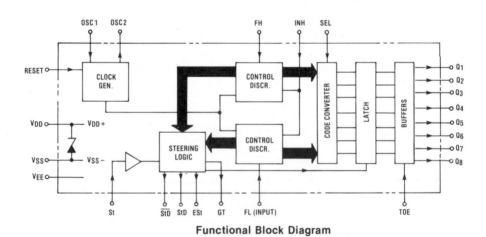

Functional Block Diagram

Fig.5 DTMF matrix (top) generates dual tones for repeater control. The touchtone pad generates two audio tones when button is pressed. For example, when "6" is pressed, tones of 1477 and 770 Hz are generated. The DTMF IC (bottom) at the repeater detects and filters tones which appear at FH and FL outputs for circuit control.

with noise components and mixing signals. Normal operation of a VHF repeater is severely hampered under these circumstances. Locating the repeater receiver outside the area of high signal density permits unrestricted reception, while locating the repeater transmitter in the downtown area provides a strong signal for mobile stations.

Repeater Activation

When a signal appears at the input channel (Fig. 4) the basic repeater turns itself on automatically by means of a **carrier operated relay** (COR). The transmitter is then modulated by the received signal. A time delay circuit holds the transmitter on the air for a few seconds after the incoming signal disappears so that quick, deep fades (**flutter**) in the received signal will not make the repeater pop on and off the air. Automatic level control circuits prevent the repeater from being overloaded by strong local signals. This allows any operator within radio range to use it.

Some repeaters require a multitone audio sequence controlled by a dual-tone, multifrequency (DTMF) device, and a **touch tone** matrix as shown in Fig. 5 to activate and control the repeater. The correct keying sequence is supposed to be known only by the station trustee or control operator of the repeater group. A variety of tones may be used for different control purposes. The technique is also used on some open repeaters when it is desired to deactivate the repeater during periods of inactivity.

In many cases, for a number of reasons, the repeater requires a coded signal before it can be activated. If the decoder in the repeater does not detect a special tone or tone sequence transmitted on the incoming signal, it does not switch to the retransmit mode. A name for one such system is **Continuous Tone Coded Squelch System** (CTCSS). When this scheme is used, a CTCSS encoder must be incorporated in the transceiver in order to activate the repeater. Encoder tones usually lie in the band of audio frequencies between 67 and 204 Hz.

The CTCSS system allows closer geographical spacing of repeaters using the same frequency than would otherwise be possible, reducing co-channel interference during band openings. In addition, it prevents false keying of the repeater by noise or by users of a nearby repeater operating on the reverse frequency pair. It is also often used to prevent uninvited operators from using the system. Some repeater control systems may be accessed by a simple whistle into the microphone, while others require complex, pulsed codes.

The Making of a Repeater

In reality, an amateur repeater consists of two principal parts. The first is the hardware and the engineering concept that creates the system, including treatment of such complex items as power source, receiver passband, operating frequencies, site location, antennas, duplexers, etc.

The second part consists of people (radio amateurs) and the manner in which they control the repeater. This part is the great unknown. The repeater users are made up of those who created the repeater, other local users, and mobiles from other areas driving through the region.

Once an amateur joins the FM activity on a VHF band, he encounters a world alien to the time-honored HF world of amateur operation. This is because an FM repeater has many of the aspects of a broadcasting system. It radiates signals from a favorable takeoff point and in many cases has a large, captive audience. Since FM receivers are channelized and have a squelch control, the receiver can be left on stand-by with the assurance that any signal that unlocks the squelch can be clearly heard. Much repeater communication, therefore, is heard by non-amateurs (family members, for example) who are "captive", either in the radio operating room or in the automobile carrying the equipment. VHF scanning receivers, moreover, are popular with non-amateur listeners and bring a large audience of eavesdroppers to amateur repeaters.

Thus, a large listening audience surrounds the repeater, and transmissions passing through the repeater are heard by parties unknown. Because of the nature of the system and the audience, the repeater user bears a heavy responsibility for his conduct on the air. Remarks uttered by anyone are heard by all and the repeater user must conduct himself in the best interests of amateur radio. Unfortunately, as in real life, there are misfits who flaunt the rules of repeater usage, sometimes leading to deliberate jamming of a repeater. Many repeater owners are adept at radio location by direction finding techniques and jammers are eventually identified and turned over to the authorities.

The Repeater Group

Because of the cost, complexity and maintenance problems associated with a repeater, most are operated by a club or association.

Repeater design, installation, operation and maintenance may be complex and costly, but actual repeater use is simple. All stations in the group transmit on the repeater input channel and listen on the output channel. It is not unusual for mobile stations only a few miles apart to communicate via a repeater that is many miles from either of them, over a radio path that will not maintain direct station-to-station contact.

There are two basic categories of repeaters: open and closed. The open repeater is for the benefit of all who wish to use it, whereas the closed repeater is designed to restrict its control to a specific group of amateurs. The term "closed" does not imply special electronic access, as many closed repeaters operate with normal carrier-operated transmit functions, just as a club net might operate on 75 meters. Some "open" repeaters may use special access methods to combat interference from another repeater on the same channel group or to allow access to regional outlying receivers. The terms, therefore, apply more to the operational than to the technical aspect of the system.

There are many more open repeaters than closed. Some repeaters fall into both categories, as they are closed but may be opened by a simple signal such as a tone, or whistling into the microphone. These repeaters are closed to prevent distant, unwanted signals from turning the repeater on.

The open repeater has great advantages to the amateur on the road and to the general public. Route information, emergency communication, aid in the case of accidents—all are available to the user of the open repeater. Amateurs visitng a new area may be able to strike up friendships with other amateurs using a local repeater. Thus the open repeater is a ready tool to enhance the amateur's traditional role of providing emergency communications when other services fail, and is far more valuable to the general amateur service than the closed version.

Using the Repeater

VHF FM is a different concept of amateur operation compared to the HF bands. Gone is the concept of calling CQ as all listeners on the channel hear you as soon as you activate the repeater by pressing your microphone button. A simple remark that you are monitoring the channel often brings a response. Phonetics are infrequently used, often leading to confusion of call letters. More experienced amateurs use alphabet phonetics to avoid errors.

The greatest sin (aside from intentional interference) on any repeater is being long-winded. Don't talk too long! Many repeaters have timers which shut the repeater down after a few minutes. Operators who monopolize a repeater are termed **alligator operators**—all mouth and no ears!

The knowledgeable repeater user waits a few seconds before replying on the repeater to allow someone else a chance to break in on a clear channel. Another operator may have priority or emergency traffic and may not be able to cut in if the repeater user comes back instantly, as is done on the lower frequency bands. Remember, the stronger signal captures the repeater and masks out the weaker signal.

Finally, don't key the repeater unnecessarily. There's a great temptation to activate a repeater by keying your mic and seeing if the repeater responds. **Ker-chunkers** drive other repeater users crazy, so do not join this unpopular group.

Repeater Channels

Repeater operation is authorized on all VHF bands. The Federal Communications Commission does not regulate which specific frequencies a repeater should use, however this is a factor that affects others, and is of prime importance in setting up a new repeater or moving the frequency of an existing one.

Repeater channels are identified by the input/output frequency pairs. Thus, a repeater whose input frequency is 146.01 MHz and whose output frequency is 146.61 MHz is referred to as an "01-61 machine".

Coordinating Committees exist among radio amateurs concerned with allocation of repeater channels for a given area. The choice of channel separation, repeater input and output frequencies and location must be made carefully. Consideration of other activity in the area must be made to prevent interference to or by the repeater. Coordination allows an amateur from one area to use his equipment on commonly accepted repeater channels in another area.

The repeater situation is continually in a state of flux, and in some areas unusual "split" and "inverted" channels are in use. It can be confusing. The prospective repeater user should check with the local radio clubs, and also obtain a copy of the "Repeater Directory", published by the American Radio Relay League. In addition, the A.R.R.L. sponsors a Repeater Advisory Committee on national VHF band planning.

Some amateurs operate simplex on the output frequency of a repeater, or listen on the input frequency and transmit on the output, thus allowing them to talk to other amateurs without actually going through the repeater themselves. This can cause confusion, with one station in contact being heard by all through the repeater, and the other one only heard by a few!

Repeater channels are usually found on the different bands as follows:

On 6 meters, the repeater input/output frequencies are separated by 1 MHz, with the channels located above 52.01 MHz. Channel spacing is 20 kHz.

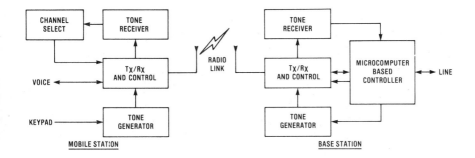

Fig.6 *Autopatch (automatic phone patch) is popular function of many repeaters. Mobile operator transmits coded audio signals which disable transmitter portion of the repeater and connects receiver portion to existing telephone circuit. Once patch has been activated, the mobile operator is able to dial the desired telephone number. When conversation is ended, control tones revert the system to normal repeater status.*

The 2 meter band presents a complex repeater picture. The repeater input/output frequencies are separated by 600 KHz but the channel spacing may be 60, 20 or 15 Khz. In addition, the pairs may be either low frequency input and high frequency output or the reverse, depending upon local agreement. Repeater congestion is severe in some areas, with repeater-to-repeater interference common. Check the Repeater Directory for information on your area.

220 MHz band repeater input/output channels are separated by 1.6 MHz and are located above 222.32 MHz. Channel spacing is 20 kHz.

On the 440 MHz band, repeater input/output channels are separated by 5 MHz, with 25 kHz spacing between the channels, starting at 442 MHz.

Repeater channels in the 902 MHz band start at 907-910 MHz (inputs) with outputs in the 919-922 MHz range.

Repeater channels in the 1240 MHz band have input/output separation of 12 MHz, with 25 kHz spacing between channels. Inputs range from 1270 to 1276 MHz, with outputs falling in the 1282 to 1288 MHz range.

Autopatch

The FM repeater may be used in conjunction with a telephone line interconnection called an **autopatch** (Fig. 6).To activate such a system, the operator needss a DTMF (**touch tone**) pad on his transceiver to create the necessary timed control tones for switching operations of the telephone circuit (Fig. 5). The autopatch connection at the repeater is linked to an existing telephone circuit. The user can deactivate the transmitter portion of the repeater and connect the receiver portion to the telephone circuit. Once the autopatch is activated, the user can dial the desired telephone number.

Sophisticated repeater systems provide the users with optional functions actuated by tone control. The repeater may be commanded to change frequency, change antennas, shut down, or transmit telemetry information back to the control operator. These support systems are limited only by the ingenuity of the equipment designer and by certain legal restrictions.

Repeater DX

Sometimes repeater operators will be able to work DX when the repeater is within a tropo duct and rebroadcasts a distant station into the repeater's local service area. The DX station may be working his local repeater and not even be aware that his signals are being rebroadcast.

DX openings may be great sport for some, but a source of interference for others, since a good opening will produce heavy interference on simplex and repeater channels. Because of this, the FM mode is not likely to produce the kind of long distance contacts that can be accomplished on CW or SSB.

Future Repeater Operation

As repeater techniques are continually being modified and advanced, the concept of VHF repeater operation poses serious challenges and offers great opportunities to the Amateur Radio Service. The next few years will test the ability of the amateurs to use and control repeaters in the best public interest.

Chapter 4

VHF MOONBOUNCE COMMUNICATION

Line-of-Sight at its Ultimate

Among the more impressive feats in the history of amateur radio are the experiments in VHF moon-reflected communication. Fig. 1 helps to visualize the enormous obstacles that had to be overcome before two-way, earth-moon-earth (EME) communication was possible. Today, properly-equipped VHF amateurs located at any two places on earth where the moon may be seen at the same time can communicate by reflecting their "line-of-sight" signals off the moon. Unlike earthbound communication where a signal takes just over a tenth of a second to circle the earth, the moon-bounce signal takes 2.38 seconds, or more, to make the round trip. This is plenty of time to hear your own signal coming back at you!

The Earth-Moon-Earth Circuit

The moon is approximately 2160 miles in diameter and orbits the earth at a distance that varies from 221,463 to 252,710 miles. An orbit takes 29-1/2 days and because the orbit is somewhat eccentric, the moon travels across a different segment of the sky each night of the lunar month. Although the moon looks big when it is full, it subtends an arc of only half a degree when seen from earth (Fig. 2). Even the highest gain amateur VHF antenna has a beam pattern much greater than this, consequently only a minute portion of the signal aimed at the moon actually strikes it, the rest passing

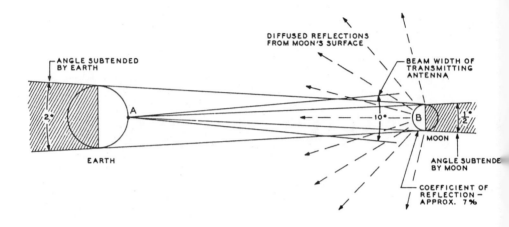

Fig.1   *The Earth-Moon-Earth (EME) VHF radio circuit. This illustration shows the great obstacles which would seem to make detection of moon-reflected signals highly improbable. The path to the moon and back is about half a million miles and the moon reflects only 7 percent of the radio energy striking it. The reflected energy is diffused all over the heavens and only a small portion of the energy which left the transmitter is reflected back to the earth. Finally, the largest VHF receiving antenna is only a fraction of the earth's pickup area facing the moon. In spite of these staggering difficulties, the EME radio circuit works with equipment well within the capabilities of many VHF operators.*

Fig.2   The radio reflector in the sky! Signals passing through the ionosphere reach the moon's surface and are reflected back to earth. Many radio amateurs communicate on the VHF bands via moon reflection.

out into limitless space. Furthermore, about 93 percent of the signal striking the moon is absorbed. And, as our astronauts verified, the moon surface is exceedingly rough, so that the 7 percent of the radiated signal that is reflected back from the moon is diffused all over space.

Viewed from the moon, the earth subtends an arc of about 2 degrees, and the VHF signal that returns to the earth is spread over an earth surface of nearly one hundred million square miles. Clearly, only a small fraction of the transmitted signal radiated to the moon and reflected back impinges upon the receiving antenna. The path loss varies with frequency and ranges from about 180 dB at 50 MHz to 207 dB at 1296 MHz. Even so, hundreds of amateurs enjoy EME communication and on a VHF DX contest weekend, literally hundreds of EME contacts take place between amateurs all over the world.

EME and the Operating Frequency

Since time, effort and money must be expended to achieve moonbounce communication, it is important that the enthusiast choose the optimum band for his activities. An error of choice at the beginning of the game could turn out to be expensive and time consuming. Several interlocking factors

Fig.3 *Moonbounce array of W6PO is poised for new DX contact. The 160 element beam is composed of 32 separate beams, aligned into eight 20 element collinear assemblies, stacked four wide. Array is mounted on a horizontal boom, 30 feet long. Overall array is 33 feet wide and 24 feet high. During the first year of operation, W6PO contacted Canada, Germany, Sweden, Hawaii and Australia.*

are involved in this decision. Moonbounce contacts have been made on all amateur VHF bands but the prospective "moonbouncer" should limit his work to a single band at first.

To begin with, the power of a radio signal travelling through space is attenuated as the square of the ratio of the frequency. The reason is that two identical antennas for different frequency bands have different receiving **capture areas**, and the higher the frequency, the less the capture area. Consequently, for two identical antennas for 144 MHz and 430 MHz, the overall "path loss" from transmitter to the moon and back to the receiver is 8.3 times (9 dB) greater on 430 MHz than it is on 144 MHz. A similar increase in path loss occurs between the 430 and the 1250 MHz band. In addition, transmitting efficiency tends to decrease, and receiver noise figure and coax line loss increase, with increasing frequency. Thus there are compelling reasons to use as low an operating frequency as possible in EME work.

On the other hand, the power gain of a directive antenna array of a given size increases by the same ratio that the path loss increases. Because antenna gain is realized in both transmission and reception at each end of the circuit, there is a net signal gain with increase in frequency, even after deducting the increased circuit losses.

Which VHF Band to Use?

At present, the various gain and loss factors seem to combine most favorably on the 144 and 420 MHz bands. Conventional power tubes work well at these frequencies and low noise receivers are easy to obtain. In addition, conventional multi-element Yagi antennas of reasonable size are relatively easy to assemble and provide sufficient gain for successful moonbounce operation.

The 50 MHz and 220 MHz bands are not world-wide amateur assignments and relatively little EME work is conducted at these frequencies. Also, certain areas in the U.S. and Canada are blanketed with "spillover" from adjacent TV channel 13, making low-signal reception difficult in the 220 MHz region.

The 1250 MHz band has sparse EME activity because of the problem of generating sufficient power in this frequency range. Conventional power tubes do not work this high in frequency and the specialized tubes that do are expensive and hard to obtain. The usual solution is to use 100 watt planar triode tubes in combination. A large dish antenna helps overcome the handicap of low power.

As a result, most amateur EME activity has taken place in the 2 meter band. Antennas for successful moonbounce work, while large, are not impossible to build and high power transmitters and low-noise receivers are commonplace among serious VHF experimenters (Fig. 3). Of great importance, too, is the high level of world-wide activity in this band and the fact that the 144-146 MHz portion is an international segment. Moonbouncers all over the world are active in this frequency region.

How Much Power is Needed?

The required transmitter power for successful moonbounce communication can be approximated from Fig.4. The scale at the right-hand side is labelled "Total Antenna Gain". The example shows that if a 144 MHz sta-

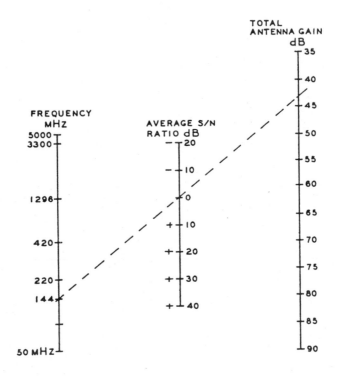

Fig.4   *Moonbounce nomograph provides guideline to successful EME contacts. The graph is based upon 590 watts transmitter output, zero dB receiver noise figure and 100 Hz bandwidth. Lay a straight edge across any two columns and read the desired unknown from the third column. The antenna gain shown represents a compromise between calculated gain required based upon free space losses and the experience of successful moonbounce operators. At 144 MHz, for example, for an average signal-to-noise ratio of zero dB, a total antenna gain (for both ends of the path) is about 42 dB. Two 22 dB gain antennas should do the job. Below 220 MHz, sky temperature changes make zero dB noise figure impractical.*

Fig.5 *Modest moonbounce array of W7JRG is compos-ed of four KLM 14 element beams. Homemade rotor is mounted atop a ten foot pole. Moonbounce array need not be high in the air, it only needs a clear shot at the moon.*

tion at one end of the EME path is equipped with an antenna with a gain of 26 dB over a dipole, the station at the other end will need only 17 dB antenna gain. The graph is based upon a transmitter power output of 590 watts (equivalent to one kilowatt input), zero decibel receiver noise, and 100 Hz receiver bandwidth. In real life, with a 17 dB antenna, some stations run more power (up to the legal limit) and have a narrower receiver bandwidth, thus providing a better circuit than illustrated in the graph.

In many instances, amateurs can take advantage of earth reflection to pick up an additional few decibels of signal gain when using the setting or rising moon. And stations having smaller antennas (Fig. 5) can often work other stations whose large antenna arrays make up the "missing gain".

Faraday Rotation

When a radio signal is radiated from the earth to the moon, it may rotate in polarization several times before it strikes the moon. And when the signal is reflected back to earth, any such rotation will be reversed on the return journey. This is called **Faraday Rotation** and is produced by the effect of the earth's magnetic field on the signal. Faraday rotation may cause a horizontally polarized signal to return to earth with vertical polarization, or with some intermediate value of polarization, because the path length between the earth and the moon is continually changing.

Faraday Rotation produces a cyclical fade in the signal which is quite slow at the lower frequencies, typically, 5 minutes at 50 MHz and 20 minutes at 144 MHz. The period of fade increases with frequency until it ceases to be significant at 1250 MHz. Special antennas that emit and receive circularly polarized waves are a partial solution, but most amateurs accept the fade on 2 meters and work around it, as the fade period is rather long, typically 20 minutes between signal peaks.

Libration Fading

Libration fading is caused by signal scattering from the rough surface of the moon, combined with a slight "wobble" in the movement of the moon. This produces a fluttery, irregular fade that can be very deep. It is most pronounced on the higher bands (i.e., above 420 MHz). On a weak signal, all that can be heard are the peaks of the fade in the form of short bursts, or "pings". Using slow speed CW, libration fading can break up the dashes and eliminate dots, making the signal audible but unintelligible.

Doppler Shift

Because the moon moves toward or away from the earth at speeds up to 980 miles per hour, **Doppler Shift** changes the frequency of the received signal from the moon compared to the transmitter frequency. Frequency shift is minimum when the moon is perpendicular to the observer, and increases with operating frequency. Maximum shift is less than 500 Hz at 144 MHz.

What Does a Moonbounce Signal Sound Like?

It takes slightly over two seconds to make the trip to the moon and back so the return echoes of your transmission can be easily received. The best way of testing an EME circuit, in fact, is to listen to the return echoes from your transmitter. It is quite an eerie feeling to send a series of dots and hear them bounce back at you a short time later! Voice signals returning from the moon have a hollow quality about them that is instantly recognizable to the moonbounce enthusiast.

Sun and Sky Noise

While general background noise in the VHF region is low on the earth, **sky noise** is an important limiting factor in reception. Most of the sky is relatively "quiet" as far as noise is concerned, but there are areas of galactic noise that can inhibit moonbounce communication when the moon passes between the earth and these areas. Sun noise is another deterrent. The sun is a prolific source of radio noise, and smart moonbounce operators avoid aiming their antenna at the sun unless they wish to use its noise to align their antenna, or other equipment.

Tracking the Moon

The moon is a moving target and several factors are involved in aiming an antenna at the moon, especially at times when it cannot be seen visually. The best way of aiming the antenna at the moon is to reference antenna direction to true North, true Zenith, the equatorial plane, and the horizon.

True North can be determined within one degree in the Northern Hemisphere by sighting the North Star. The latitude and longitude of your station can be obtained from city engineers, surveyors, airport maps and

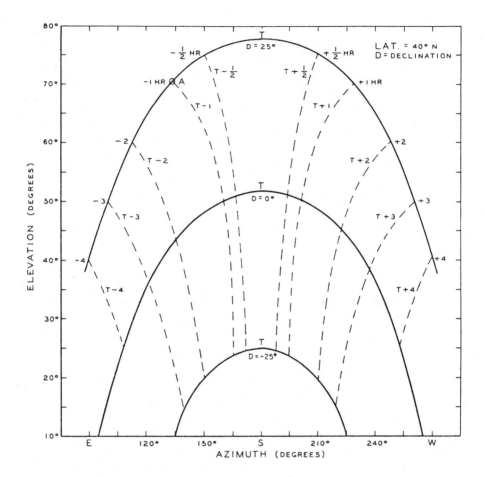

Fig.6 *Moon position from the center of the U.S. may be determined with this chart. The altitude of the moon above the horizon and its azimuth change minute by minute, every day, but they repeat each lunar month. The "Nautical Almanac" and similar manuals predict the moon position far in advance.*

This graph is plotted for a position of 40 degrees north, with the observer facing south. At time T the moon is due south of the observer. At T-1 (one hour before T), the moon is at point A. Given the date, the declination and the local time that the moon appears due south, the azimuth and elevation can be found. Curves for other values of declination can be interpolated and drawn in between these three curves. (Reference: Lund, "How High the Moon?", QST, July, 1965.)

local radio and television stations.

The true Zenith (straight up) must also be known. A plumb line dropped from a step ladder or other support to within a few inches of the ground can determine this. A stake is then driven in the earth parallel to the plumb line. The horizon is 90 degrees from the Zenith in all directions.

Finally, the equatorial plane (90 degrees from true North) is determined by driving a second stake into the ground,at an angle above the horizon to the south equal to your latitude. It will point to the equatorial plane.

Once these directions are fixed, you can point your antenna at any object in the sky, aided by published almanacs and sky maps.

The next problem is to determine the position of the moon. As indicated in Fig. 6, the moon's position in the sky is continually changing. Published charts predict its position for every hour of the day far into the future. The "Nautical Almanac", available from the U.S. Government Printing Office, has convenient charts and is recommended. With this data at hand, the moonbounce antenna can be aimed without even seeing the moon.

The Moon Position

Moon positions are listed in the "Nautical Almanac" for each hour of the day in **Coordinated Universal Time**(UTC) in terms of the **Greenwich Hour Angle**(GHA) and the **declination angle** (Fig. 7). The GHA is defined as the angle in degrees to the west of the meridian at Greenwich, England. If the GHA is 20 degrees, for example, the moon is directly over the 20 degree west longitude meridian. The declination angle indicates how far above or below the equatorial plane the moon will be at this time.

The GHA is continually changing because of the change in orbital velocity of the moon and the earth's rotation.

The declination angle of the moon defines one part of its orbit. During a month, the moon progresses through a range of north and south declination angles. The angle must be converted to provide an elevation angle at the altitude of the observer.

If the moonbounce antenna is supported on a **polar mount**, such as often used for telescopes, setting the mount to the local hour angle and the local declination automatically aims it at the moon, and a motor driven mount will track the moon all evening.

If the moonbounce antenna is supported on an azimuth-elevation mount (one that is positioned in azimuth—the compass direction, and elevation—the angle above the horizon), the Nautical Almanac and the Hydrographic Office

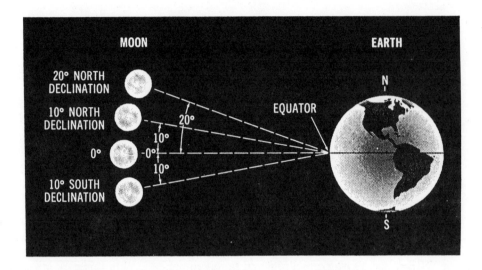

Fig.7 *Declination angle of the moon. During a month, the moon progresses through a range of northern and southern declinations.*

Publication HO-214 for your latitude are required to determine the moon position. Both publications are available from the Government Printing Office, Washington, DC 20401. Various computer programs in BASIC, FORTRAN, and RPN that are useful in determining azimuth and elevation of the moon are listed in the "ARRL Handbook". In some cases, amateurs are using personal computers to automatically control the antenna heading.

Aiming the Moonbounce Antenna

The moonbounce antenna requires a sighting tube or **boresight** in order to visually aim it accurately at the moon. This can be a small-diameter aluminum tube two feet long. The alignment can be accomplished by aiming the antenna at the sun, peaking it for maximum noise. The boresight tube is next aligned on the sun and firmly clamped to the antenna structure. The experimenter should never look through the boresight tube because of possible eye damage from solar radiation. A symmetrical spot of sunlight cast upon a piece of paper held at the back end of the tube will indicate proper alignment.

Fig.8   Backyard moonbounce array of W7FN is made up of sixteen Yagi arrays. Antenna elevation and azimuth are adjusted by hand.

Equipment Requirements for EME Operation

What does the experimenter need in the way of station equipment for moonbounce operation? The EME path is marginal at best and maximum legal transmitter power is required for long-term operation. Users of medium power equipment (500 watts or so) with antenna gain in the range of 17-18 dB have had successful EME contacts, but they are only able to contact the more powerful stations having the best and largest antennas.

A **low-noise** receiver is mandatory for EME work. On the 2 meter band, a noise figure (NF) of 0.5 dB is suggested, as the receiver is then limited only by cosmic noise (assuming coax line loss is low). The newer GaAsFET preamplifiers will do the job at 144 and 432 MHz.

In order to overcome the considerable path loss, large antenna arrays are required at both ends of the EME circuit for serious EME work. Many stations use four long Yagis, each of which provides a power gain of about 14 dB. Total array gain is about 20 dB gain over a dipole. Some stations combine four groups of four into a sixteen-antenna array that provides an additional 6 dB power gain (Fig. 8).

The more antenna gain one station has, the less gain is required for another station to work it, everything else being equal. That is why many beginning moonbouncers can work other EME stations while using only one or two high gain Yagi antennas—the other stations are providing the missing antenna gain!

Finally, regardless of antenna gain and excellence of equipment, the best EME station can be crippled by transmission line loss. On transmit, valuable power is lost in the line, and on receive the line loss is a source of noise which deteriorates the overall noise figure of the system. Some operators overcome this handicap by using rigid, air-insulated coax line, while others mount a preamplifier directly at the antenna. If the distance from antenna to the station is not great, foam-filled RG-8/U type coax may be used with good results.

The EME Reporting System

While some EME stations run high power, have big antennas and correspondingly big signals, most EME signals are very close to the noise level. For this reason, a special reporting system has been devised to provide quick and reliable confirmation of a valid contact. The system is composed of dashes only, as dots have a low energy content and tend to disappear in the noise. The TMO system (popular on the 2 meter band) is as follows:

T- The letter T is sent repeatedly when the signal can be heard but no intelligence can be detected.

M- The letter M is sent repeatedly when portions of call letters can be copied.

O- The letter O is sent when a complete call set is copied (SM7BAE DE W6PO is considered a call set.)

R- The letter R is sent when an O report has been copied.

SK- The standard sign-off is sent signifying that the O and R reports have been copied.

RST- The standard RST system is used if the signals are loud enough.

Chapter 5

AMATEUR SATELLITE COMMUNICATION

From Sputnik to OSCAR

Radio amateurs have been interested in space communication since the first Sputnik beacon sent telemetry signals back to earth on 20 and 40 MHz in 1957. A few experimenters seriously discussed the idea of an amateur space beacon and finally in late 1959 hams in California formed Project OSCAR (Orbital Satellite Carrying Amateur Radio), a group dedicated to building and launching an amateur space satellite. The challenge was enormous and the task formidable, but within two years a suitable beacon and power supply had been built and tested by the group (Fig. 1). Approval for launch was received from the U.S. Air Force and, on a space-available basis, the first Oscar satellite was launched in late 1961.

Successful orbit of the tiny satellite was confirmed by KC4USB in Antarctica and before the historic flight had ended, the 140 milliwatt, 145 MHz beacon had been logged by over 570 amateurs in 28 countries in all continents. Amateur radio had achieved its place in space history!

Oscar 3, a Milestone in Amateur Radio

After the launch of a second beacon satellite, Project Oscar designed a new and startling concept: a multiple-access communication satellite,

Fig.1 First Oscar satellite was launched in 1961 and con-
tained a low power 2 meter beacon. The 140 milliwatt signal
was heard by amateurs worldwide, and over 570 tracking
reports were received on the flight. The satellite was built
by Project Oscar radio amateurs in California.

available for use by radio amateurs in all countries on an equal basis.

This pioneer satellite carried a tracking beacon and a repeater which
received and retransmitted a band of frequencies in the 144 MHz range
(Fig. 2). Unlike an FM repeater which retransmits a single channel, the
Oscar repeater retransmitted all signals within a 50 kHz spectrum. During its short life, Oscar 3 set new VHF records on virtually every pass
around the earth. Amateurs W1BU and DL3YBA established the first trans-
Atlantic satellite contact via Oscar 3 and a whole new world of international, long distance VHF communication was at hand.

Various successful Oscar satellites, prepared by Project OSCAR and
AMSAT (Radio Amateur Satellite Corporation) in cooperation with amateurs
in many countries, were built and launched in the following decades. These
were soon to be joined by amateur satellites built by amateurs in the U.S.S.R.
and launched by the Soviet space agency. As of mid-1987, 24 amateur
spacecraft have been successfully placed in orbit, and they have logged
a total of more than 50 years of active service!

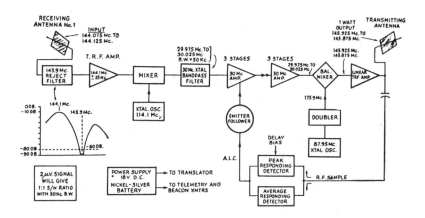

Fig.2 Block diagram of Oscar 3 frequency translator system. Received signals feed into bandpass filter, followed by an RF amplifier and mixer stages. The 30 MHz IF signal is amplified and mixed back to 2 meters. Linear amplifier stage provides one watt PEP output signal. Automatic level control circuit reduces effect of overload from strong signals.

Early satellites operated in the 2 meter band, but later satellites also made use of various HF beacons, plus the 28, 420 ,1250 and 2400 MHz bands. Two Soviet satellites had robot "stations" aboard which would automatically work CW QSOs with amateurs on earth! Later satellites carried packet radio and interesting digital experiments.

Satellite Communication For You

From time to time new amateur satellites are launched into space. Depending upon the operating modes of the "birds", they often may be heard on simple receiving equipment as they speed around the world. To use a satellite to work other amateurs, however, is a more complicated task requiring knowledge of the satellite orbit and use of some specialized communication equipment. Getting into the satellite "game" is not difficult, but it does require learning a few new techniques. Observing an amateur satellite ground station in action is the best way to get a "feel for the action."

Generally speaking, an amateur communication satellite is launched into a nearly-circular orbit or an elliptical orbit about the earth. The satellite remains in this orbit while the earth rotates within it. As the earth rotates, most areas of it will pass beneath the satellite.

Amateur satellites are divided into two groups: Phase II (low altitude) and Phase III (high altitude) spacecraft. (The original Phase I satellites are obsolete.) Each satellite in each group has individual transmitting and receiving modes (including CW, SSB, RTTY, packet, etc.) and arrangements of **up-link** (satellite receiving) and **down-link** (satellite transmitting) frequency ranges. Some of the various frequency modes are:

> A- 2 meters up, 10 meters down
> B- 70 cm up, 2 meters down
> J- 2 meters up, 70 cm down
> K- 15 meters up, 10 meters down
> L- 24 cm up, 70 cm down
> S- 70 cm up, 13 cm down
> T- 15 meters up, 2 meters down
> KA- 15 and 2 meters up, 10 meters down
> KT- 15 meters up, 10 and 2 meters down

Thus, you must decide which satellites and modes your station will be designed for, as it is complex and expensive to design a station that will work with all combinations of frequency and mode. Such a decision should take into account the satellites presently operational as well as satellites planned for future launch.

The Linear Repeater

Unlike regular FM repeaters, the repeater (or **transponder**) aboard an amateur satellite is a linear device; that is, it faithfully reproduces CW, SSB, RTTY and other types of transmissions. And, unlike regular FM repeaters, the satellite repeater covers not just a single channel but a range of frequencies called the **passband**, which may cover 100 kHz, or more. Since the satellite retransmits every signal it receives in the passband, many stations can use the satellite trasponder simultaneously.

The Satellite Beacons

In addition to the communication equipment, each satellite carries one or more beacons which transmit telemetry to the monitoring stations. The

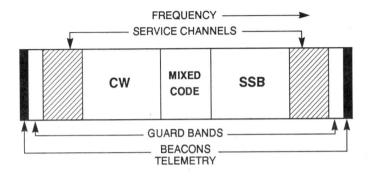

Fig.3 *A representative satellite passband. Beacon and telemetry signals are at the edges of the passband. CW and SSB signals occupy the ends of the translator portion of the passband. In some cases, the downlink passband is a mirror image of the uplink band, that is, signals at the lower edge of the input band appear at the upper edge of the output band. There is little standardization among satellites and users must follow instructions carefully to make the satellite respond properly.*

easiest way to locate a satellite is to monitor the beacon frequency.

It is common for a satellite to carry equipment to work more than one frequency mode, and new combinations of up/down links may be designated as a new mode. Some satellite transponders invert the uplink signal, that is, change upper sideband on the uplink to lower sideband on the downlink, or change an operator's increase in uplink frequency to a decrease in downlink frequency.

The satellite passband is generally broken up into small frequency segments reserved for beacons, telemetry and control, and larger segments for general usage (CW, SSB, RTTY, etc.). This passband information is published for each satellite (Fig. 3).

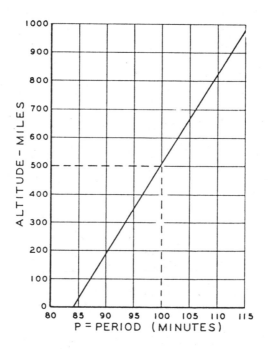

Fig.4   Orbital period of a satellite is related to the altitude. For example, if the "bird" has a period of 100 minutes, the altitude is 500 statute miles for a circular orbit. The higher the satellite, the farther away you should be able to hear it. Satellite quickly plunges back to earth if the period is much less than 90 minutes.

Tracking the Elusive Satellite

Unless the satellite is placed in a **geostationary orbit** (in which it appears that the satellite hovers above a specific location on the earth's equator) it is necessary for the experimenter to know the location, flight path, and **orbital period** (time for one revolution about the earth).

The practical question to be answered is: When will the "bird" be in radio range and where should my antenna be pointed? The satellite is a moving target and aiming information must be constantly updated.

One simple tracking technique is to monitor the beacon frquency with an omnidirectional antenna, such as a ground plane. When the satellite comes within range, the beacon will be heard. The operator can then switch to a beam antenna and aim it for maximum signal strength.

The ability to track a satellite, however, adds enjoyment and enhanced operator capability to the game. Basic information to help you in this regard is at hand. Orbital predictions are broadcast and otherwise publicized by the ARRL, AMSAT and others before and during a flight. The information is provided in a short format, mainly for those using simple map-based tracking calculators for low altitude satellites; a more complete format is available for those using the microcomputer tracking programs.

The orbital information provided in the short format includes the predicted times of satellite south-to-north **equatorial crossings**, (given in UTC); and the points of crossings in degrees of west longitude. The time difference betwen successive crossings provides the orbital period (Fig.4) and the difference in equator crossing point longitudes provides the **longitude increment** to the west during an orbit. All of these parameters are usually predicted with good accuracy ahead of launch time. Fortunately, the job of keeping track of an amateur satellite is not difficult and can be achieved with a minimum of figuring and fuss if the proper data is at hand.

The complete format includes a description of the satellite orbit, the orientation of the orbital plane, and the position of the satellite on its flight path at a specific time. This collection of constants is referred to as the **orbital elements**.

Low Altitude Satellites

Low altitude satellites are usually launched into a circular orbit about the earth. Orbital height and period are related to each other. If, for example, the orbit of the satellite is 500 miles above the earth, the orbital period will be 100 minutes. If the period is as low as 85 minutes, the satellite will not remain in orbit, but will plunge back to earth.

Once the satellite's height has been determined from the orbital period, the maximum **access distance** (range from a point on earth beneath the satellite along the surface of the earth to the satellite's horizon) may be determined (Fig. 5). At an altitude of 500 miles, for example, the ground range is nearly 2000 miles. Two experimenters who are on opposite sides of the satellite's visbility circle (about 4000 miles apart) could theoretically communicate with each other through a repeater satellite orbiting at that height.

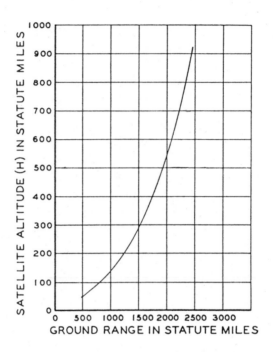

Fig.5 *Orbital altitude versus maximum access distance.*
This is the distance measured along the ground between
the point directly below the satellite and a station who sees
the satellite at the horizon.

The Low Altitude Satellite Path

A clear picture of the whereabouts of a low altitude satellite is gained by visualizing the satellite as rotating about the earth in a fixed, invisible plane, with the earth revolving inside the satellite orbit. As the earth rotates within the orbit, all areas of the world will pass beneath the satellite, except for areas near the north and south poles, which will be missed if the orbital plane is tipped with respect to the earth's axis (**inclination**).

Using this imaginary picture, a scale model of a typical low altitude satellite pattern may be made up with the aid of a small globe and a loop of wire to simulate the orbit plane, as shown in Fig. 6. The globe rotates within the ring, and the satellite travels along the ring as it circles the earth. Thus, when the satellite passes over a particular spot on earth on one or-

Fig.6 *Model of low altitude orbit satellite path. The or-bit is made of stiff wire, tilted with respect to the axis of the globe by the amount of orbital inclination of the "bird". If, for example, the inclination is 80 degrees, the wire orbit just reaches the 80 degree latitude circle, as shown here.*

bit, the rotation of the earth causes the satellite to pass over a different spot, lying to the west of the first spot, on the next orbit (**longitude increment**). Each successive orbit progressively crosses the equator of the earth farther west and, to an earthbound observer, each successive orbit has moved further west from his point of observation.

In reality, the observer has moved east with the earth's rotation, and the orbit of the satellite has remained fixed in the heavens. Half a day later, the observer's position has rotated 180 degrees (12 hours) and he is looking at the reverse side of the satellite orbit. If he started out watching north-to-south passes, he is now watching south-to-north passes! These motions of the satellite and the earth are summarized in Figs. 7 and 8.

Calculating the position of an amateur satellite from raw data may be an interesting exercise in spherical geometry and trigonometry to those so inclined, but much of the pencil work can be avoided by using data in this chapter in conjunction with the printed data and informational broadcasts provided by various tracking nets and by the ARRL headquarters station, W1AW. Programs for several small computers are also available from AMSAT.

Once the satellite is "found", the orbital predictions may be confirmed by listening to the satellite beacon for a few successive passes.

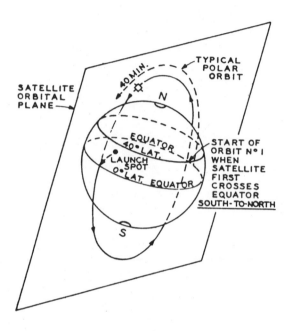

Fig.7   *The earth rotates within satellite orbit, which lies in an orbital plane. As the earth rotates, all areas of the world will pass beneath the satellite with the exception of areas near the north and south poles which are missed if the orbital plane is tilted with respect to the axis of the earth. Orbit number one starts when a satellite first crosses the equator in a south-to-north direction.*

Using a Low Altitude Satellite—An Example

A low altitude, VHF repeater satellite opens up frequencies not normally used for very long distance communication. Since the uplink/downlink (input/output) frequencies of a forthcoming satellite are not known until a few months before launch, specific operating data for new satellites cannot be provided here. However, a program based on the early Oscar 6 satellite can give you a general picture of low altitude satellite communication.

Oscar 6 was a repeater satellite that "listened" on the 2 meter band

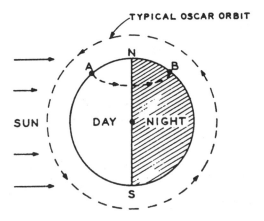

Fig.8   *The satellite orbit remains nearly fixed in space while the earth revolves inside it. If, for example, a satellite is launched from California in a southward direction in daylight, daytime passes for the next few weeks will be in a a north-south direction and nighttime passes will be in a south-north direction. For daytime passes, the observer is at point A and for nighttime passes, at point B.*

uplink and retransmitted the received signals in a narrow downlink passband in the 10 meter band. The satellite accepted CW and SSB signals. For example, if you transmitted to the satellite on 146.00 MHz, your signal would be retransmitted back to earth on 29.55 MHz.

It is easy to determine if your signal reached the satellite: listen on the down link frequency and see if you can hear your own signal coming back! This technique allows you to adjust your antenna position and power output for maximum efficiency.

Once the satellite has been found by listening for the beacon, the downlink of the satellite should be monitored to familiarize yourself with the operating techniques. Signals coming out of the satellite will not all be the same strength, thus the station with an effective receiving antenna will be the most successful in "pulling" signals out of the satellite. The types of receiving and transmitting antennas depend, of course, upon the frequencies used by the satellite.

The polarization of signals repeated from the bird constantly changes due to satellite tumble and Faraday Rotation (see the previous chapter). Severe fading can be a problem. The most successful experimenters use crossed Yagi beams for circular polarization to minimize fading.

Once you hear your own signal coming back through the satellite, and other amateurs hear you, you are well on the way to working DX. Operating range of the satellite is known before launch, as is the length of time it will be within range of your area. The amount of time you can use a low altitude satellite varies with the launch parameters, orbit, and your latitude. For Oscar 6, a typical "DX opening" lasted up to about 20 minutes and stations within a circle of about 2500 miles radius centered on a point on earth directly below the satellite could work each other.

Doppler shift (a downward shift in frequency due to the motion of the satellite) is present on signals received from low altitude satellites and you will have to follow the received signal as it slowly changes frequency. No doubt about it, operator expertise is required, and the best way to gain expertise is by on the air experience!

Using a High, Elliptical-orbit Satellite

Some satellites are launched into long elliptical orbits in which the height constantly changes. The orbit **apogee** (point farthest from the earth) can be as much as 23,000 miles distant, and the **perigee** (point closest to the earth) can be as little as 300 miles. This orbit takes the satellite a great distance out into space where it is in view of nearly half of the earth for the greater portion of a day. By comparison a low altitude satellite is only in view of roughly one-tenth of the earth, and is in view at any earth point for only a short time. The high altitude satellite provides reliable, predictable communication for a long period of time for ground stations, without requiring elaborate tracking antennas.

A representative orbit of a Phase III satellite is shown in Fig. 9. Special maps, tables and computer programs are available for use with these interesting satellites. The useful "window" of the satellite is as it approaches and leaves the apogee. The orbit of the satellite progresses across the surface of the earth, as does the orbit of a low altitude satellite, and the experimenter must follow the changing ground track of the "bird" from orbit to orbit. But once the orbit is determined, the ground station antenna usually can remain fixed for extended periods of tme, often more than one hour, during the portion of the satellite orbit that "illuminates" the area of the ground station.

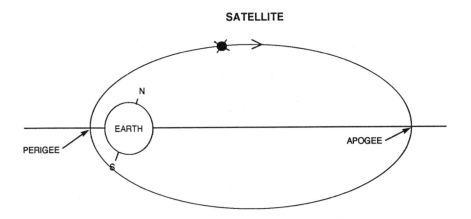

Fig.9 *Representation of elliptical orbit. The velocity of the satellite depends upon position in orbit. Near the apogee, the satellite moves relatively slow. The orbital plane of the satellite is referenced to the equatorial plane of the earth. The inclination of the orbit is the angle between the axis of the earth and a perpendicular to the orbital plane.*

The High, Geostationary Orbit Satellite

The **geostationary** orbit is a circular orbit directly over the equator at a height of about 23,100 miles. If the satellite is carefully launched eastward with an inclination of zero into a circular orbit of this type, its angular velocity will equal that of the earth. To an experimenter on the earth, the satellite appears to remain motionless in the sky.

This orbit makes the satellite ideal for certain types of radio communications. Orbital tracking is unnecessary and Doppler shift is absent. Commercial communication satellites are "parked" along the geostationary orbit and used for television and communication relay points.

A variation of the geostationary orbit is the **synchronous orbit**. In the most common case, the satellite does not remain in constant position above a point on the equator, but moves back and forth around the point in a figure-8 ground track. This orbital pattern is useful for certain types of commercial and military satellites. Up to mid-1987, neither of these orbits have been put to use in the Amateur Service due to lack of launch opportunities and the complexity of the satellite systems. However, plans are underway to produce a geostationary amateur satellite (Phase IV A and B) for launch in the early 1990's.

Additional Information

Things change so rapidly in the world of satellite communication that the prospective user requires a constant update on activities. For additional information on amateur satellites the reader is urged to read QST magazine and "The Satellite Experimenter's Handbook", by M. Davidoff, published by ARRL, 225 Main St., Newington, CT 06lll. For the latest news and information, the "Amateur Satellite Report" can be obtained from AM-SAT (Radio Amateur Satellite Corp.), Box 27, Washington, DC 20044. The weekly 80 meter AMSAT nets are held on Tuesday evenings at 8 p.m. (local time) and have been a prime source of information for over 15 years. The current net frequency is 3840 kHz.

Chapter 6

ALMOST EVERYTHING ABOUT COAXIAL LINES

Your "Radio Hose"

Your antenna is a device that changes electric waves flowing in a conductor into electromagnetic waves in space or vice versa. Radio energy flows to or is taken from the antenna via a **transmission line**, or "radio hose" (Fig. 1). Before antennas are discussed it is a good idea to examine transmission lines and determine how they perform in the VHF region. One determining factor in this matter is the amount of power lost in the line. This is a function of the loss per foot of the line and the total line length. Some lines are more lossy than others, wasting valuable RF, and the VHF enthusiast should use the best line he can buy. This chapter gives some guidelines in this regard.

The Coax Line-Good and Bad

Modern VHF equipment uses **coaxial line** (abbreviated **coax**) for a conductor to the antenna system. Plenty of acceptable coax is available, along with some very poor cut-rate cable made for the CB market. The problem is separating the good stuff from the junk.

Most coax lines in amateur use are identified by the manufacturer's type number and also by a military serial number. For a given line made by several manufacturers the type number may change but (if the line is military-approved) the military number remains the same. The military

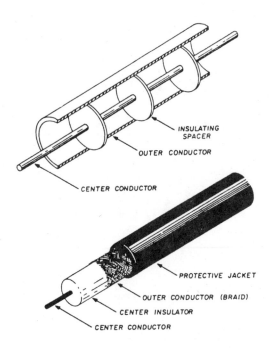

Fig.1   *The coaxial cable is a "radio hose" through which radio energy flows to and from your antenna. A representative coax line consists of a center conductor surrounded by a low-loss insulating jacket and a braided or solid copper outer conductor which acts as a shield. The cable is covered by a waterproof vinyl jacket extruded over the outer conductor. The solid outer conductor coax is called "hardline" (top). If the coax has a braided outer shield, it is flexible enough to be bent around a radius greater than twenty times the diameter of the cable.*

designation can be recognized by the prefix "RG" followed by a serial number. A typical example is cable identified by "RG-8/U", the "U" indicating utility-type coax for general use.

The problem lies in the fact that the military designation "RG-8/U" for 52-ohm coax was dropped many years ago when the Armed Forces switched from a 52-ohm to a 50-ohm system. This opened the door for the manufacture and sale of questionable coax bearing a seemingly authentic military acceptance number.

Any outfit owning a second-hand cable-making machine can grind out cheap "RG-8/U"-type cable made to any specification it desires, and an unsuspecting buyer may be conned into thinking he's getting a high grade product. Not so! Some RG-8A/U (50-ohm) is still made to military specifications, but the great majority of RG-8/U is not.

The less expensive cable looks good to the casual observer but unless it is made by a reputable manufacturer it may have less than the optimum number of fine wires in the outer braid, or perhaps the inner conductor is made of smaller gauge wire. The inner insulation may be reclaimed material, the outer jacket may have pin holes in it, or alignment between inner conductor and shield may be imperfect. All of these factors contribute to cable loss.

This stuff may work OK at low frequencies (say, 160 or 80 meters) but because even the best of flexible coax is relatively lossy in the VHF region, the VHF operator should buy the best coax he can afford.

Coax Lines for VHF Operation

The most important consideration in choosing coax for VHF operation is the efficiency of the line. Generally speaking, large diameter coax lines have less loss than small diameter ones, and lines having a tightly woven outer braid (97 percent coverage) have less loss than those having a loosely woven braid (70 percent coverage).

A list of some lines suitable for VHF work is given in Table 1. Line loss is tabulated as a percentage of power lost per 100 feet of line for the popular VHF bands. Line loss is directly proportional to length, and increases logarithmically with frequency. Thus, a 10 foot (3.05 m) length of line has only one-tenth the loss of a 100 foot length at a given frequency.

For short coax runs—20 to 30 feet—at power levels up to 100 watts output on the 50, 144 and 220 MHz bands, small-diameter RG-8(M), RG-58C/U or RG-58(F) are suggested. For longer runs, and for work at 450 MHz, RG-213/U or RG-8A/U should be used. The latter cables can

CABLE	ATTENUATION (dB/100 ft.)					POWER RATING (WATTS)				
	50 Mhz	144 Mhz	220 MHz	420 MHz	1240 MHz	50 MHz	144 MHz	220 Mhz	420 Mhz	1240 MHz
RG-58C/U	3.0	6.0	8.0	15.0	23.0	350	175	125	90	40
RG-58 (F)	2.2	5.4	6.9	10.1	15.1	400	200	150	100	70
RG-8A/U RG-213/U	1.5	2.5	3.5	5.0	9.1	1500	800	650	400	180
RG-8 (F)	1.2	2.1	2.5	3.5	6.5	1500	800	650	400	180
9913	0.64	1.4	1.9	2.7	4.7	400	200	150	100	45
RG-8 (M)	2.5	5.7	7.1	10.3	16.0	800	400	300	200	140

Table 1 Coax cables for VHF use. RG-8A/U and RG-213/U are recommended for high power and medium-length cable runs. For short runs and ease of installation, the smaller diameter RG-8(M) or RG-58C/U are suggested. Foam dielectric cable is indicated by (M). Power ratings shown are absolute maximum for unity SWR. Ratings for general amateur service are about half of these figures. The small-diameter RG-8 type coax is termed RG-8/X by Amphenol and RG-8/M by Saxton and Radio Shack. It requires a UG-176/U adapter to fit a PL-259 plug.

also be used at 1250 MHz, if the user remembers that he will lose about 17 percent of his power for every ten feet of line. And, he'll lose more than half his power (3.36 dB) in a 30 foot line.

Foam-filled coax has slightly less loss than an equivalent line having a solid inner dielectric. This may be an advantage, but the foam-style line should not be bent sharply as the inner conductor will eventually move about in warm weather when it is placed under stress. Losses will increase and ultimately, the inner conductor may short to the shield at the bend point.

For somewhat better efficiency, **Belden 9913**, or equivalent, coax is suggested. It has a solid center conductor, semi-solid dielectric, and a double outer shield. At 1250 MHz, a 50 foot length has a loss of 2.8 dB. Special solid jacket cables are also available which have even lower losses than the 9913 type.

Coax Plugs and Receptacles

The so-called "UHF-type" connector pairs used with much VHF communication equipment are the PL-259 plug and SO-239 receptacle. They are for use with cable having an outer diameter of 0.405" such as RG-8A/U and RG-213/U. With a reduction adapter (UG-175/U) the PL-259 is also usable with small diameter coax, such as RG-58A/U or RG-58C/U. These plugs and fittings are fine for the HF spectrum, and are satisfactory for amateur use through the 2 meter band at power levels up to 500 watts, and up to 450 MHz at reduced power.   Unfortunately, they are not waterproof and allow moisture to enter the end of a line if the fitting is exposed to rain. At higher frequencies and high power levels these fittings should be avoided. The newer and better type-N family of connectors are waterproof and are much preferred. At 1250 MHz, use of these coax fittings is mandatory.

The N-family includes the popular UG-21B/U plug and the UG-58/U receptacle. Also available are a large number of cable splicers, adapters and reducers.

The BNC-family of connectors is designed for the smaller diameter RG-58A/U and RG-58C/U coax. This includes the UG-88/U plug and the UG-290/U receptacle. These fittings are useful in low power equipment up to 450 MHz.

Placing coax plugs on a line is an art, not a science. If you go about it properly, it isn't hard at all. If you do it improperly, you could be in big trouble! Let's make sure you do it the right way.

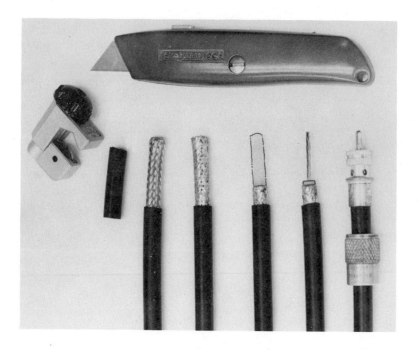

Fig.2 *Coax onto PL-259 plug—the easy way. Midget tubing cutter (left) and utility knife (top) are used. At left is a sample coax line with the outer jacket removed by the knife. Next, the outer, braided shield of the line is tinned. Third view shows the shield cut to length by the tubing cutter. Fourth view shows inner insulation cut to length and inner conductor tinned. Right view shows coax plug and ferrule on the line, with plug in position for soldering to shield thru four holes in plug shell. Gun or iron with high wattage and small tip is recommended.*

Installing The PL-259 Plug

The widely available PL-259 is for use with RG-8A/U or RG-213/U coax. Fig. 2 illustrates the necessary installation steps.

1. The first step in preparing the coax for the plug is to slide the coupling ring of the plug onto the cable with the ring threads toward the open end of the cable. Next, take a sharp utility knife (**Stanley 99A Shop Knife**, for example) and circumscribe a cut in the outer black jacket of the line. Make the cut at right angles to the cable and about 1-1/2 inches in from the end of the cable. Try not to nick the braid. The small cylinder of jacket material you have cut is now slit from one end to the other and removed.

2. You have now exposed over an inch of the shiny, outer copper braid of the cable. Without disturbing the braid, which should be lying flat against the inner insulation, take a small soldering iron or gun and quickly tin the exposed braid, rotating the cable as you go along.

With a little experience, you can make the braid into a solid entity. Don't overheat the braid, or the inner insulation will melt and seep out through the basketweave strands of the braid. (CB-type coax with 70 percent or less braid coverage will do this every time!) Clean the left-over flux from the braid with a rag moistened with paint thinner or alcohol.

3. The next step is to trim the soldered braid to the correct length. Use a small tubing cutter to do the job. The **General Hardware Midget Tubing Cutter** is recommended. Cut the braid so that one-half inch is left on the cable end. To do this, mark a line this distance back from the black vinyl jacket and place the tubing cutter over the braid, letting the cutting wheel fall on the mark. Tighten the cutter slightly and slowly revolve it about the cable, continuing to slowly tighten it as you turn. After four or five turns the cutter will have neatly cut the soldered braid. The unwanted slug of braid may then be snipped off with a pair of wire cutters.

4. Now, trim the center insulation. Cut it cleanly with the utility knife so that a collar 1/16-inch wide extends beyond the soldered braid. Don't nick the center conductor. Once the insulation is cut, pull it off the cable by grasping it and gently pulling, rotating it at the same time so it follows the twist of the inner conductor wires. When the slug is off, tin the center conductor.

5. The cable end is now ready for the PL-259 plug. Push it carefully on the cable end, rotating it so that the internal threads of the plug screw onto the outer vinyl jacket of the cable. Make sure the inner conductor is centered in the plug pin. As the plug body is screwed onto the cable, you will see the tinned braid appear in the four solder holes in the shell. Continue twisting the plug until the braid is completely visible through

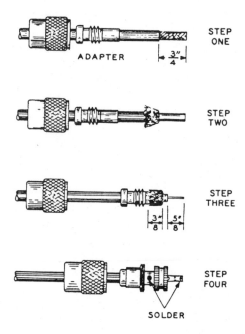

Fig.3 *Assembly technique for use with PL-259 plug and UG-175/U or UG-176/U adapter to match small-diameter cable such as RG-8/M or RG-58C/U.*

all holes and the center conductor is visible at the end of the center pin.

6. The plug is now fixed on the cable by soldering the braid through the four solder holes. Use a soldering gun with a small tip and proceed with care, using small diameter solder. Hold the plug in a bench vise during this operation. Take care that solder does not run over the outer threads of the body. The last step is to solder the center conductor to the tip of the pin and trim it. After the assembly cools down, slide the coupling ring down and screw it onto the plug.

Note: It takes longer to read about this than it does to do it! After working with one or two plugs, you can make this termination a work of art.

Using the PL-259 Plug With Small Coax Cable

The PL-259 plug can be used with small cable (RG-58A/U, for example) by adding an appropriate reduction adapter (type UG-175/U for RG-58C/U) as shown in Fig. 3.

1. The end of the cable is passed through the coupling ring and the adapter, with the threads of the ring and the narrow end of the adapter facing the end of the cable. Using the utility knife, cut 3/4-inch of the vinyl jacket off the cable. Fan the braid out slightly and carefully fold it back evenly over the adapter. Next, trim the braid with small scissors to about 3/8-inch length, so that it fits about the barrel of the adapter.

2. Now, take the utility knife and remove 5/8-inch of the insulation from the center conductor. This will leave a 1/8-inch collar beyond the braid. When the slug of insulation has been removed, tin the conductor.

3. Carefully screw the plug body onto the adapter. The center conductor of the cable should pass easily through the center pin of the plug, and the strands of the braid should appear through the side holes of the shell. Using a small soldering gun, solder the braid through the holes. Lastly, solder the center conductor to the pin and slide the coupling ring down over the plug. Finished!

Installing the Type-N Plug

The N-family of coax fittings are designed to be used with 0.405 inch diameter cable (RG-8A/U or RG-213/U) up through 1250 MHz. The connectors are considered waterproof under normal conditions. The following assembly sequence is recommended (Fig. 4):

1. Remove 9/16-inch of the outer vinyl jacket of the coax with a utility knife, as described earlier. The shiny, copper braid is then carefully combed out, using a large pin or small pointed instrument, and then folded back upon the cable. Next, cut the inner insulation with a utility knife or razor blade, removing 7/32-inch of material. The center conductor is now tinned.

2. Carefully smooth out the copper outer braid and taper it over the tinned conductor so that you can slide the nut, washer, gasket and clamp over the vinyl jacket of the line. Smooth the braid back over the clamp and trim the ends flush with sharp, small scissors. Solder the center conductor to the pin with a small iron or gun. Avoid excess heat and solder. If necessary, trim excess solder away with a small file and then clean the butt-end of the inner insulation. Make sure the center contact is flush against the insulation.

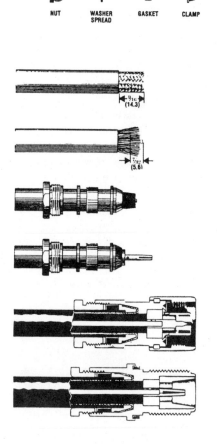

NUT   WASHER   GASKET   CLAMP   FEMALE   JACK BODY   MALE   PLUG BODY
           SPREAD   CONTACT   CONTACT

Remove $^9/_{16}$"(14.3mm) of vinyl jacket. When using double shielded cable, remove $^5/_8$"(15.9mm).

Comb out copper braid as shown. Cut off dielectric $^7/_{32}$"(5.6) from end. Tin center conductor.

Taper braid as shown. Slide nut, washer and gasket over vinyl jacket. Slide clamp over braid with internal shoulder of clamp flush against end of vinyl jacket. When assembling connectors with gland, be sure knife edge is toward end of cable and groove in gasket is toward the gland.

Smooth braid back over clamp and trim. Soft solder contact to center conductor. Avoid use of excessive heat and solder. See that end of dielectric is clean. Contact must be flush against dielectric. Outside of contact must be free of solder.

Slide body into place carefully so that contact enters hole in insulator. Face of dielectric must be flush against insulator. Slide completed assembly into body by pushing nut. When nut is in place, tighten with wrenches. In connectors with gland, knife-edge should cut gasket in half by tightening sufficiently.

NOTE: For armored cable slide cap over armor first. Push armor and cap back out of way and proceed with assembly as directed above using armor clamp in place of standard clamp nut. When assembly is complete straighten bulge in armor and trim so it can be clamped between nut and cap.

Fig.4   Assembly sequence for type-N waterproof coax plug. Shown here is the UG-21/U (Amphenol 82-96) used with the RG-8/U family of cable, Matching cable receptacle is UG-238/U (82-63). The straight adapter between two plugs is UG-29B/U (82-l0l). Drawing courtesy Amphenol Corp.

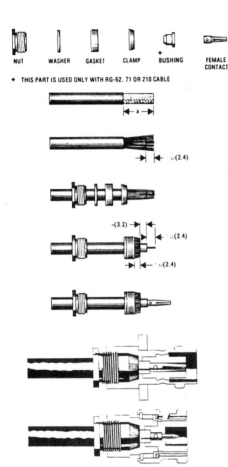

NUT   WASHER   GASKET   CLAMP   BUSHING   FEMALE CONTACT   JACK BODY   MALE CONTACT   PLUG BODY

✦ THIS PART IS USED ONLY WITH RG-62, 71 OR 210 CABLE

Cut jacket to correct dimension. Select part number or cable from table below:

Part number or RG-/U cable	Dimensions inches(mm)
P/N 31-1, -2, -18, -20 -23, 14525. RG-55, 71, 142	$5/16$(7.9)
P/N 31-12, -21, -11. RG-59, 62, 210	$19/64$(7.5)
RG-58, 140, 141	$9/32$(7.1)

Fray shield and strip inner dielectric $3/32''$(2.4). Tin center conductor.

Taper braid and slide nut, washer, gasket and clamp over braid. Clamp is inserted so that its inner shoulder fits squarely against end of cable jacket.

With clamp in place, comb out braid, fold back smooth as shown and trim $3/32''$(2.4) from end.

Slip contact in place, butt against dielectric and solder. Remove excess solder from outside of contact. Be sure cable dielectric is not heated excessively and swollen so as to prevent dielectric from entering into connector body.

Push assembly into body as far as it will go. Slide nut into body and screw in place with wrench until tight. For this operation, hold cable and shell rigid and rotate nut.

Fig.5 *Instruction sequence for BNC-type coax plug. This twist-on plug is for use with small-diameter coax. Amphenol part number is 31-002 and matching receptacle is 31-210. Drawing courtesy Amphenol Corp.*

3. Finally, slide the plug body into place so that the center pin enters the hole in the center insulator. Slide the completed assembly inside the plug body by pushing the retaining nut. When all is in place, tighten with small wrenches. When completed, the center pin conductor can be felt by pasing the ball of the thumb across the end of the plug.

Installing the Type BNC Plug

Assembly of the type BNC plug is shown in Fig. 5. The following sequence is recommended:

1. For use with RG-58-type coax, the outer insulation is cut back 9/32-inch. The wires of the outer shield are unbraided and combed out evenly around the cable, as shown in the top two illustrations.

2. After nut, washer, gasket and clamp are passed over the braid, the braid is combed out again and folded back over the tapered clamp. It is then trimmed as shown.

3. Next, the center pin is slipped in place so that it butts up against the dielectric. It is now quickly soldered with a small iron or gun. Clean the pin with a small file, if necessary. The body of the plug is now screwed into place. The nut is tightened to the plug body with pliers to complete the assembly.

Understanding the Coax Line and SWR

Transmission lines have a property called **characteristic impedance**. The value is expressed in ohms, and is determined by the physical and electrical configuration of the line. The size of the conductors, the spacing between them and the insulating material (dielectric) all contribute to the impedance figure. Most coax used by American amateurs has an impedance value of 50 ohms. Electrically, this figure expresses the ratio of voltage to current in the line when the line feeds a resistive load whose value is equal to the line impedance. It has nothing to do with the dc resistance of the line, which is quite low.

Under the above condition, all of the energy put into the line is dissipated in the load, less the amount lost because the line is not 100 percent efficient.

A transmission line terminated in this manner is said to be **matched**. If, however, the line is terminated by a load that is not equal to the impedance of the line, a portion of the power put into the line is reflected back from the load to the input end.

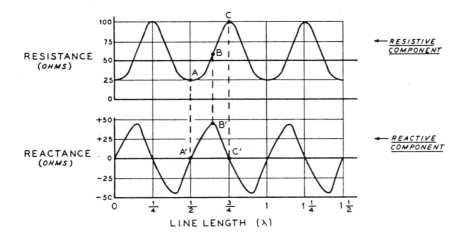

Fig.6 *The input impedance of a mismatched coax line changes with the amount of mismatch and the length of the line, even though the SWR on the line remains constant. In this example, feedpoint resistance of the antenna is 25 ohms (at left) and the SWR is 2-to-1. The feedpoint value is repeated one-half wavelength down the line (points A-A'). But at B-B' and C-C' the resistive and reactive components are widely different than at the input end of the line. Trimming the line won't change the SWR, but it might make a more satisfactory match to the output circuit of the transmitter.*

A physical analogy of this situation is the reflection of a water wave when it hits an obstacle—a smaller wave is created that is reflected back toward the source of the wave. The forward-travelling wave (the **incident wave**) and the **reflected wave** interact with each other and form interesting patterns of interference on the surface of the water.

In the case of the transmission line, the reflected power interacts with the outgoing power so as to produce points of maximum and minimum current and voltage on the line. The result is that resistive and reactive values along the line vary as a function of line length. The line now has a **standing wave** upon it (Fig. 6). The reflected power is eventually dissipated by the terminating load on the line (the antenna), but only after it has made one or more trips back and forth along the transmission line. Thus, there can be power flowing in two directions along the line if the terminating load is not equal to the characteristic impedance of the line.

The Standing Wave Ratio (SWR)

The ratio of the maximum to the minimum value of voltage along a transmission line is called the **standing wave ratio** (SWR or VSWR). The SWR is an indication of the degree of mismatch between the antenna load and the impedance of the line. The SWR is always expressed as a number greater than one (unity). A 50 ohm coax line terminated with an antenna having a 50 ohm feedpoint has an SWR of 1-to-1 (sometimes written 1/1 or 1:1).

A line does not have to be any particular length to have a standing wave on it. The only requirement for the presence of a standing wave is that reflection occurs at some point along the line. The usual reasons for a reflection are improper line termination or a faulty coax plug or connection.

The SWR is defined by the following expression:

Standing Wave Ratio (SWR)= Zo/Zt or Zt/Zo

where Zo is the characteristic impedance of the transmission line (usually 50 ohms) and Zt is the load resistance of the antenna.

The greater the difference between Zo and Zt, the larger the SWR reading. Thus, if a 50 ohm line is terminated by a 25 ohm antenna load, the SWR is 50/25, or 2/1 (two-to-one). Likewise, if a 50 ohm line is terminated by a 100 ohm antenna load, the SWR is 100/50, or 2/1. Values of SWR may be read by an **SWR meter**, which is discussed in a later chapter.

Off-frequency Operation of the Antenna

Most antennas are designed for a single frequency but may be used at frequencies slightly removed from the design frequency. As an example, many 2 meter Yagi beams are designed at a center frequency of 146 MHz but are operable over the range of 144 to 148 MHz. The design frequency is called the **resonant frequency** and at this frequency the antenna presents a pure resistive load to the transmission line. If the transmitter is operated at a different frequency (145 MHz, for example), there will be an appreciable value of SWR on the transmission line even though the antenna still has a feedpoint resistance of close to 50 ohms. Off-frequency operation of the antenna adds another factor to the matching problem: **reactance**. The reactance may have either a positive or negative value.

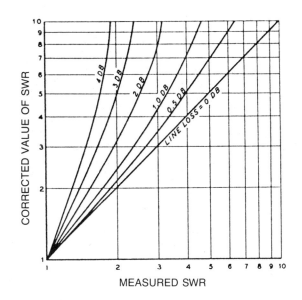

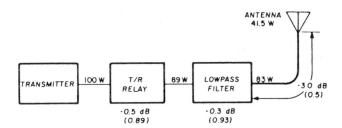

Fig.7 *RF loss in coax rises with high value of SWR (above). Circuit losses add up, as shown below. In this VHF example, 100 watt transmitter output shrinks to 41.5 watts at the antenna due to total circuit loss of 3.8 dB. The reduction factor for each loss is shown in parenthesis).*

Effects of High SWR on the Coax Line

There is nothing wrong with antenna reactance. All antennas exhibit it to a greater or lesser degree. It could be ignored except for the fact that some transmitters "don't like" an antenna load that has a large value of reactance, or that is not close to the 50 ohm nominal value of load resistance. Many VHF transceivers have a protective circuit that gradually reduces the power output of the amplifier stage as the antenna load departs from a true 50 ohm value. If the SWR is excessively high, the transmitter turns itself off. In addition, the greater the departure from SWR = 1/1, the greater the loss in the transmission line (Fig. 7).

Transmission Line Rules

Without belaboring the point, and without going into the theory of transmission lines or a study of standing waves on the line, the following two simple rules summarize what makes an antenna and feed system work efficiently:

1. The antenna must be resonant at some frequency within the operating band being used.

2. The resonant antenna should present to the line (directly or through a **matching device** at the antenna) a feedpoint resistance roughly equal in ohms to the characteristic impedance of the transmission line. The matching device may be a balun, gamma match or similar unit.

Chapter 7

VHF Vertical and Mobile Antennas You Can Build

In most areas of the world amateur VHF FM is vertically polarized. That is, the active antenna elements are vertical with respect to the earth's surface. Vertical polarization started out many years ago because it was easier to mount a vertical antenna on a vehicle than any other type. And since much FM work concerns mobile operation, vertical polarization remains popular today.

The "Rubber Ducky" Antenna

What type of antenna does your HT (hand-held transceiver) use? The answer, of course, is a short, flexible vinyl covered whip, fondly referred to as a **rubber ducky** by humorists. This is an approximate quarter wavelength of wire wound into a short helix and covered with an insulating jacket. Compared to a full-size quarter-wave whip it is quite inefficient, but it does the job for short-range communication. On most hand-helds the rubber ducky antenna snaps onto a connector, and is easily removed.

Build A Quarter-wave Whip

You can extend the range of your HT by substituting a quarter-wave whip for the rubber ducky antenna. Full-size whips with a coax connector can be purchased, or you can build your own, as shown in Fig. 1.

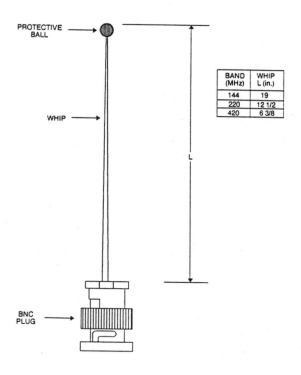

BAND (MHz)	WHIP L (in.)
144	19
220	12 1/2
420	6 3/8

Fig.1   *Full size whip for your HT extends operating range. The whip is made from flexible brass or hard-drawn copper rod. Lower end is soldered into pin of BNC-type plug (UG-88C/U, or equivalent). Antenna is held in place by epoxy cement poured into plug. No connection is made to shell of the plug.*

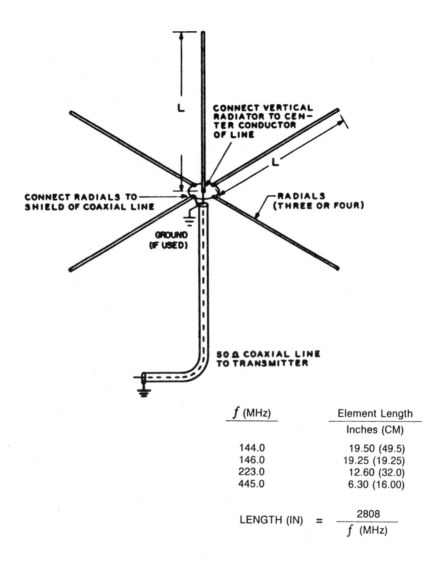

ƒ (MHz)	Element Length
	Inches (CM)
144.0	19.50 (49.5)
146.0	19.25 (19.25)
223.0	12.60 (32.0)
445.0	6.30 (16.00)

$$\text{LENGTH (IN)} = \frac{2808}{ƒ \text{ (MHz)}}$$

Fig.2 *The VHF ground plane antenna. Tubing is used for the 50 and 144 MHz whips and no. 12 enamel coated copper wire for the higher frequency antennas. The radials can be made of either wire or tubing for the low frequency antennas. The antennas for 220 MHz and higher are so small they may be mounted on the reverse end of a PL-259 or type-N coax receptacle.*

A coax plug mating to the case connector is required. The whip is made from a flexible length of brass or hard-drawn copper rod. Cut it according to the table and solder one end into the center pin of the plug. It is held in position by pouring a bit of epoxy cement into the plug after the wire is positioned. No connection is made to the outer shell of the plug.

The free end of a whip antenna is a dangerous thing as it may damage a person's eye if it is improperly handled. Accordingly, the top quarter-inch is cleaned and a length of thin, tinned copper wire is wrapped around it to make a ball. The ball and whip top are then coated with solder to form a solid, round end to the whip.

The Ground Plane Antenna

Most VHF FM operators use a simple ground plane antenna for their home stations. This consists of a quarter-wave vertical whip mounted above three or four horizontal quarter-wave radials which constitute a "radio ground" for the antenna. The radials are slanted downwards to enhance local coverage and to improve the match to a 50-ohm line. When mounted in the clear this small, unobtrusive antenna gives good results out to the radio horizon. Many manufacturers provide inexpensive ground plane antenna kits, or you can build your own on a coax receptacle as shown in Fig. 2. The ground plane can be bolted to a TV-type chimney mount or house mount for quick installation.

An Inexpensive "Gain Antenna" for 2-Meters

This simple multi-element vertical antenna designed by Fred Dietrich, NM6J, combines good gain, low SWR and very low construction cost. Shown in Fig. 3, the omnidirectional array is only about six feet high. The antenna structure is made of a length of 3/4-inch Schedule 40 (thick wall) PVC plastic water pipe. Overall pipe length is sufficient so that the antenna can be supported by clamps at the bottom end. For this 2 meter model, an eight foot length is used. The top of the pipe will be closed with a PVC cap cemented in place after the antenna wire is installed, using the liquid sealer provided for such material. Before sealing, a small hole is drilled in the cap and the tip of the antenna wire passed through it. The tip is bent over and held in place with a spot of cement.

The radiating portion of the antenna is a no. 12 copper wire 52-1/4 inches long. About one inch of extra wire is added to this length to allow it to be affixed to the cap with PVC cement, and to form a solder connection

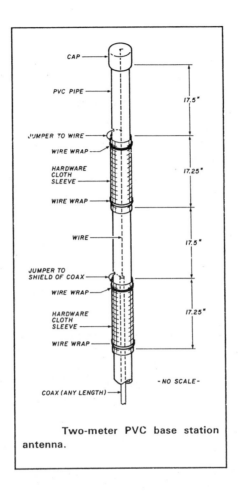

Two-meter PVC base station antenna.

Fig.3 *A stacked, gain antenna for 144 MHz. Only six feet high, this antenna provides omnidirectional gain suitable for FM service. The antenna is constructed on a length of PVC plastic pipe having an i.d. of 3/4-inch. The antenna phasing sleeves are made of hardware cloth, wrapped around the pipe and soldered in position with retaining wires. The antenna is mounted in a vertical position.*

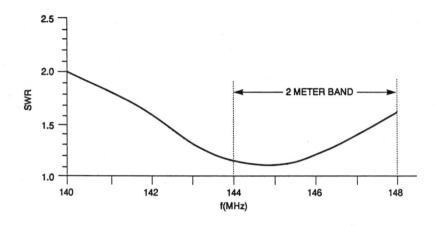

Fig.4   The SWR response across the 2 meter band of the NM6J stacked, gain antenna.

to the coax line at the base of the antenna. The wire left over is cut off during assembly to establish the resonant length.

Two **phasing sleeves** are used with the antenna. They are made of galvanized iron hardware cloth having a mesh of about 3/8 x 3/8-inches. The cloth is folded around the PVC pipe and wrapped with wire to hold it in place. Two sleeves are required. Each sleeve is 17.25 inches long. The retaining wires are soldered to the hardware cloth at several points around the circumference.

A short jumper wire joins the top of the upper sleeve to the antenna wire running inside the PVC pipe. It is suggested that this wire be soldered to the antenna wire, then fished out through a small hole drilled in the pipe. If this and the following steps are done before the top cap is fastened in place, the assembly will proceed smoothly.

A second, similar phasing sleeve is affixed to the structure below the first, as shown in the illustration. This sleeve is connected to the outer shield of the coax line by means of a short wire passed through a second hole drilled in the PVC pipe. After assembly, all holes are filled with a spot of cement making the assembly waterproof.

The antenna is mounted in a vertical position and the coax line to the station brought down vertically beneath the antenna. A representative SWR plot of antenna performance is shown in Fig. 4. NM6J has not measured the gain of the antenna, but he says it will "open up" repeaters heretofore inaccessible with a ground plane antenna previously mounted in the same position.

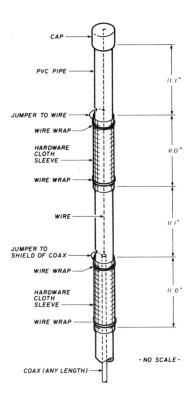

Fig. 5   *Dimensions of the NM6J stacked gain antenna for the 220 MHz band.*

A "Gain Antenna" for 220 MHz

The antenna described in the previous section can be built for 220 MHz service (Fig. 5). Note that because of the dielectric effects of the PVC pipe and end loading, the dimensions are slightly shorter than those derived by scaling the antenna down in size.

Construction details are similar to those given for the 2 meter design. (This variation of the "Gain Antenna" was devised by Tom McMullen, W1SL.)

The Expanded 144 MHz Ground Plane Antenna

A ground plane antenna whose vertical element is 5/8-wavelength long and whose radials are 3/4-wavelength long provides substantial gain over the conventional antenna having quarter wavelength elements. A practical design for an expanded ground plane is shown in Fig. 6.

The antenna is made out of brass rod, or welding rod, about 1/8-inch diameter. The whip is cut 52 inches long to allow for some pruning. The four radials are precut to a 59 inch length.

The coil form is cut from a rod of LEXAN insulation, obtainable from some hobby shops or large industrial material outlets. This is a polycarbonate material with good resistance to the ultra-violet rays in sunlight.

The form is drilled at the top to accept the whip and at the bottom in two places to attach to the mounting block.

The base coil is wound on the insulating form. It consists of 6-1/2 turns of no.14 tinned copper wire, spaced 1-1/8 inch long. The coax line is tapped on the coil about 2-1/3 turns form the bottom end. The bottom of the coil is grounded to the aluminum mount. The top end of the coil is attached to the whip by a short strap which encircles the base of the whip and is held in place with 4-40 hardware. A coaxial receptacle is fastened to the bottom of the mount and the whole antenna is mast-mounted by an L-shaped bracket.

The radials are placed in four mounting holes drilled in the side of the mount and are locked in position with 4-40 set screws. They are connected electrically to the shell of the receptacle.

The center conductor of the coax receptacle is attached to a tap point on the coil by means of an insulated wire brought out through a hole drilled in the side of the form.

To adjust the antenna, the feedline wire is tapped on the coil about two turns from the grounded end. The antenna is temporarily erected in the air, clear of nearby objects. The vertical whip is trimmed 1/4-inch at a time and the SWR response of the antenna checked across the band after each cut. The SWR should bottom out at about 1.5-to-1 at the middle of the band with a very broad null, indicating good bandwidth performance.

Squeezing and expanding the bottom coil may help to lower the SWR reading a bit. The final step is to adjust the coil tap, a quarter-turn at a time, until the point of lowest SWR is achieved near the middle of the band. When properly adjusted, the SWR response will show less than 1.6-to-1 at the band edges and about 1.1-to-1 at the center frequency.

Coil tap, number of turns on the coil and vertical antenna length are interrelated. If the antenna is too short, increasing coil inductance by

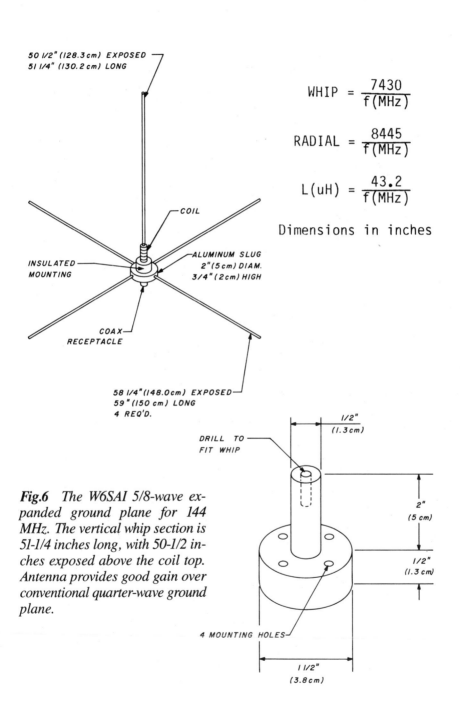

$$\text{WHIP} = \frac{7430}{f(\text{MHz})}$$

$$\text{RADIAL} = \frac{8445}{f(\text{MHz})}$$

$$L(\text{uH}) = \frac{43.2}{f(\text{MHz})}$$

Dimensions in inches

50 1/2" (128.3 cm) EXPOSED
51 1/4" (130.2 cm) LONG

COIL

INSULATED MOUNTING

ALUMINUM SLUG
2" (5 cm) DIAM.
3/4" (2 cm) HIGH

COAX RECEPTACLE

58 1/4" (148.0 cm) EXPOSED
59" (150 cm) LONG
4 REQ'D.

DRILL TO FIT WHIP

1/2"
(1.3 cm)

2"
(5 cm)

1/2"
(1.3 cm)

4 MOUNTING HOLES

1 1/2"
(3.8 cm)

Fig.6 *The W6SAI 5/8-wave expanded ground plane for 144 MHz. The vertical whip section is 51-1/4 inches long, with 50-1/2 inches exposed above the coil top. Antenna provides good gain over conventional quarter-wave ground plane.*

BAND	WHIP		RADIALS		COIL
(MHz)	IN.	CM	IN.	CM	(μ H)
28	259.00	658	294.0	746.0	1.54
50	147.6	375	167.8	426.0	0.86
144	51.25	130.2	58.25	148.0	0.30
220	33.54	85.2	38.12	96.8	0.19
440	16.77	42.6	19.1	48.5	0.09

Dimensions for the W6SAI 5/8-wave expanded ground plane for the various VHF bands. Whip length and coil inductance are adjusted for lowest SWR on feedline at chosen design frequency in band. Data shown is for center of band.

squeezing the turns together will lower the antenna resonant frequency. The antenna is very forgiving, and if an extra-eager "snip" cuts the antenna too short, a slight compression of the base coil brings the antenna back to resonance.

On-the-air tests indicate the antenna is doing what it is supposed to do: provide a good signal at a distant repeater that could not be triggered with a conventional ground plane antenna.

The Mobile Antenna

Quarter and three-eighths wave vertical whip antennas are widely used for VHF mobile operation. The ideal mounting spot is on the roof of the car, with a rear-deck mount a good alternative. For temporary operation a magnetic mount can be used, although it can slip and let the antenna blow off the vehicle at high speeds, particularly when driving into a strong wind. A more permanent installation involves attaching the antenna base directly to the vehicle. Many amateurs hesitate to do this because it involves cutting a hole in the car body.

Within a few miles of a repeater, good communication can be established by merely using a hand-held unit within the vehicle. This is of great benefit in times of an emergency. Many amateurs, in addition to a hand-held, also carry a quarter-wave whip with magnetic mount. The hand-held can be quickly attached to the emergency antenna with a short length of coax line.

A Portable J-pole Antenna for 220 MHz FM

This is an ideal portable antenna for emergency service. It provides good signal gain over a "rubber duckie" and may be used with any HT. Hang it from a tree branch, or other support, and see how your range increases!

The antenna is cut from a 40-inch length of 300 ohm TV-twinlead (Fig. 7). One end of the line is stripped back for 3/8-inch and the exposed conductors are soldered together. The line is trimmed to 38-1/4 inches, measuring from the shorted end. This end of the line is left open. A paper punch is used to make a small hole in the insulating web in order to hang the antenna with a length of string.

One of the conductors is cut 13-1/4 inches from the shorted end to leave a 1/4-inch gap. The longer section of this conductor is unconnected and the short section forms the matching stub for the antenna.

The last step is to cut a length of RG-58C/U about ten feet long. Place a matching plug on one end that fits the receptacle of your HT. At the other end, trim the outer jacket back about two inches. Using a pin, unbraid the outer shield of the line and twist it into a pigtail lead. Trim the insulation from the center conductor back to within a quarter-inch of the pigtail. Finally, expose the wire of the short conductor of the twinlead about two inches from the shorted end. Do the same for the opposite conductor. Tin these exposed points and solder the pigtail of the coax to the exposed point on the **short** conductor. Solder the inner wire of the coax to the **long** conductor, as shown in the illustration. Tape these joints to make them waterproof.

The antenna may be hung by the top from any convenient point, provided it is held clear of nearby metallic surfaces.

A Mobile J-pole Antenna for 144 MHz

An effective and inexpensive mobile or home station VHF antenna is the base-fed dipole, or **J-pole** antenna shown in Fig. 8. The antenna is a half-wavelength long, fed at the bottom with a quarter-wavelength matching stub and a coaxial balun. It is designed to be used with a 50-ohm coax line. The J-assembly is quite rugged and is mounted to a vehicle or other support by means of a clamp attached to the shorting bar, or by a window mounting block. Alternatively, the antenna can be mounted to a photo tripod for emergency use.

The feedline is attached to the antenna by means of U-shaped clamps cut from thin aluminum. The lips of the clamp are compressed by 4-40 nuts and bolts to make a tight fit to the antenna rods. A low value of SWR is achieved by moving the tap points (A-A) up and down the antenna, a quarter-inch at a time.

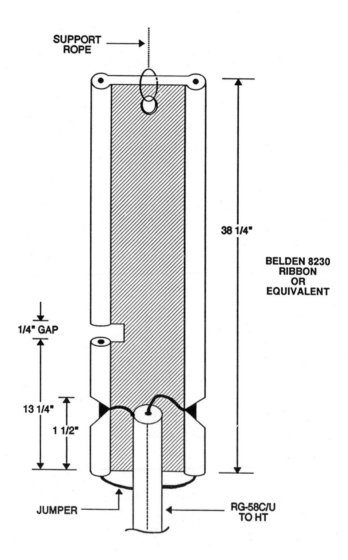

Fig.7   *A length of ribbon line makes a portable gain antenna for 220 MHz. It provides substantial improvement over a ''rubber duckie'' and may be used with any HT. Hang it from a tree branch and see how your signal range increases!*

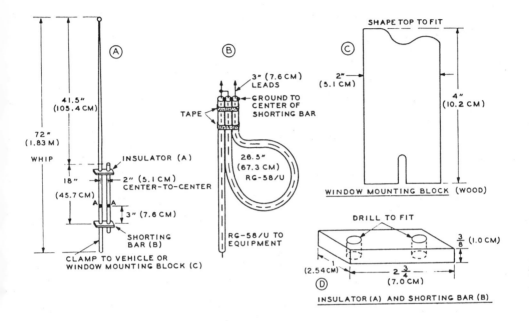

Fig.8 *A J-pole antenna for 2 meters. This popular antenna is designed to mount in the open window of a vehicle. It is made from a cut-down CB whip, with an auxiliary 1/4-inch diameter aluminum tube serving as a matching section. The tube is connected to the whip at the bottom by shorting bar B. Insulator and shorting bar are tapped for set screws to hold antenna together. The window mounting block (C) is clamped to the bottom portion of the whip below bar B for quick, easy installation on vehicle. Transmission line and balun are tapped on antenna at points A-A.*

The shorting bar is made of a piece of aluminum and the insulator is LEXAN or lucite.

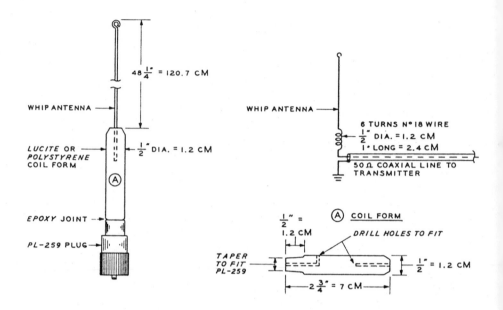

Fig.9   *An extended 5/8-wave whip for 2 meter work pro-vides improved coverage for FM operation. Whip is made from replacement CB HT antenna. The polystyrene coil form is drilled at one end to accept the antenna. The other end is drilled to pass wire connection from center pin of coax plug. The wire is fished out through the side hole and connected to loading coil wound on coil form.*

The 5/8-Wavelength Whip Antenna for 144 MHz

A less conspicuous mobile antenna than the J-pole is the base-tuned antenna shown in Fig. 9. It provides about 3 dB gain over a simple whip and when mounted on the roof, or the trunk lid of your vehicle provides a good signal in all directions. The base coil and whip length are adjusted for lowest SWR on the coax line at your chosen operating frequency.

A PL-259 plug is shown in the drawing but many HTs require a BNC-type plug. In this case, the coil form is not tapered, but is drilled out to fit over the shell of the plug. No connection is made to the shell.

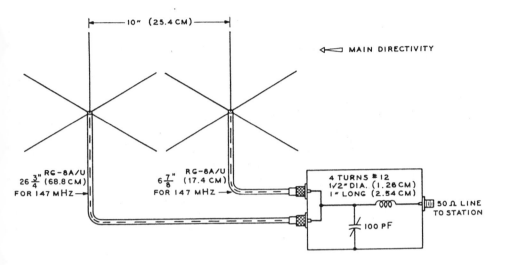

Fig.10 *A phased ground plane array is useful for fixed-station work. Beam provides broad coverage in-line with antennas. Whip and radials are 19 inches long for 147 MHz design frequency.*

A Phased Ground Plane Array for 144 MHz

The universal VHF mobile antenna is vertically polarized, especially in the world of FM, and a fixed station desiring to communicate efficiently with mobiles is forced to use a vertical antenna to avoid signal loss through cross-polarization effects.

Omnidirectional radiation and vertical polarization are usually employed for a repeater, but in some areas the best location for the repeater may be off to one side of the desired service area. Or the base station desiring to use the repeater may be in the same situation. Under such circumstances, a directional antenna may be desirable.

Figure 10 shows a simple, rugged directional antenna having a **cardioid** (heart-shaped) pattern. The array is made of two phased ground plane antennas, spaced 1/8-wavelength apart. The radiation pattern is quite broad and a power gain of about 4 dB in the favored direction over a simple whip antenna is achieved.

In this design, the phase difference between the antennas required to achieve the desired radiation pattern is established by the lengths of feedline to each antenna measured from a common point, at which the lines are connected in parallel. They are matched to a 50-ohm line by means of an L-network mounted near the antenna in a waterproof metal box. The capacitor and spacing between the turns of the coil are adjusted for lowest SWR on the coax line to the station.

The Horizontal Mobile Antenna

In the case of VHF SSB, or serious over-the-horizon DX work, the horizontal antenna array is preferred by many amateurs. For mobile work, the simple **halo** antenna is popular as it radiates a horizontally polarized signal, yet is compact enough to mount on a 4- to 6-foot vertical pipe clamped to the vehicle with a bumper mount. The halo is merely a horizontal half-wave dipole bent into a circle or V-shape, as shown in the next section.

A Horizontal Mobile Antenna for 144, 220 or 420 MHz

Shown in Fig. 11 is an inexpensive and easily constructed antenna which works well in mobile SSB operation and provides a horizontally polarized signal. The antenna consists of a V-dipole having an interior angle of about 90 degrees. It is mounted on the top of a length of 7/8-inch diameter PVC water pipe. The length of pipe is determined by the vehicle installation. If a rear bumper mount is used, the pipe should be four or five feet long so the antenna clears the top of the car body. If it is mounted on the rear deck at the hinge line of a hatch-back, the pipe can be much shorter.

The antenna consists of two lengths of brass or welding rod, about 1/8-inch in diameter. The rods pass through two close-fitting holes drilled through one end of the plastic pipe at right angles to each other and about 1/2-inch apart. The dipole elements cross over each other inside the pipe. Trim the rod length for lowest SWR on the coax line.

The easiest way to hold the rods in position is to drill out a metal washer so it fits closely over the rod. Force it onto the rod a sufficient distance so that it is flush against the PVC pipe when the rod is in position. Then solder the washer in place.

The free ends of the rods are soldered to an RG-58/AU coax line. The shield of the line goes to one rod and the inner conductor to the other. After connections are made, the end of the line is sealed to make the joint waterproof.

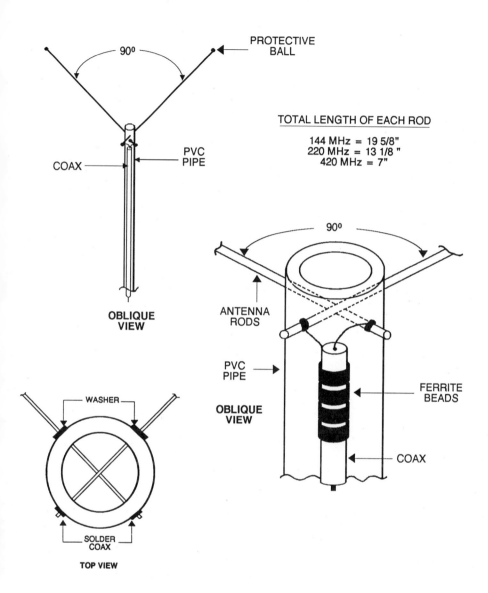

TOTAL LENGTH OF EACH ROD

144 MHz = 19 5/8"
220 MHz = 13 1/8 "
420 MHz = 7"

Fig.11 A horizontally polarized mobile antenna for 144 or 420 MHz SSB. It consists of a V-dipole supported atop a section of PVC pipe. The antenna can be attached to a rear bumper mount or placed on the rear deck of a hatch-back, depending upom the length of the supporting pipe. It may also be mounted on the roof with a magnetic mount.

In order to preserve an omnidirectional antenna pattern, it is necessary to isolate the coax shield from antenna current. This is done by slipping ten ferrite beads (Amidon FB-77-6301, or equivalent) over the coax just below the point where it attaches to the antenna. The beads are held in place with tape or heat-shrink tubing.

To protect a passerby from being injured by the antenna rods, the rod ends are cleaned and tinned. Small diameter, tinned copper wire is wound around each tip to form a ball about 3/8-inch diameter. The ball is covered with solder which affixes it to the antenna end.

If the antenna is mounted to the vehicle with a bumper mount, the elements are positioned so that one is in line with the side of the vehicle while the other one is directed inwards towards the opposite side. This prevents the rods from extending beyond the body of the vehicle.

For mounting atop the metal roof, a magnetic mount can be used, with the plastic pipe reduced to a length of 20 inches.

Chapter 8

VHF BEAM ANTENNAS YOU CAN BUILD

"Aluminum is cheap!"

—Anonymous DXer

The old adage is true: If you can't hear 'em, you can't work 'em. The secret to hearing them and to being heard lies mainly in your antenna system. VHF antennas are no exception to this old adage and newcomers quickly find that the stations with the outstanding signals are the stations with better, more efficient antenna systems.

VHF beam antennas are very attractive. They are small because the wavelength is short, they are easy to build, and they don't cost very much. Many high gain VHF arrays are available in kit form at your local distributor of ham gear.

Many amateurs, on the other hand, enjoy building their own beam antennas. There are many different types, and some of the most practical are discussed in the following sections.

Beam Antenna Terminology

You'll hear some interesting terms that apply to beam antennas. Some of these are: **directivity, power gain, beam width** and **front-to-back ratio**.

Directivity is the ability of a beam antenna to concentrate radiation, or to receive energy, in a particular direction as compared to an antenna which has no directivity. Directivity is always the same for a given antenna, whether it is being used for transmission or reception. All practical antennas exhibit directivity to a degree; in a beam, directivity is very important.

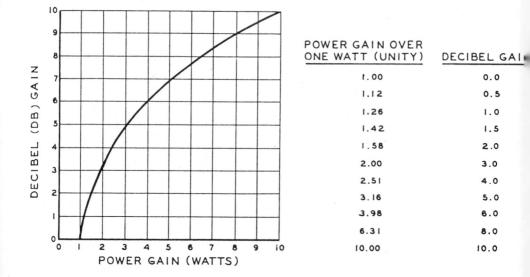

POWER GAIN OVER ONE WATT (UNITY)	DECIBEL GAIN
1.00	0.0
1.12	0.5
1.26	1.0
1.42	1.5
1.58	2.0
2.00	3.0
2.51	4.0
3.16	5.0
3.98	6.0
6.31	8.0
10.00	10.0

Fig.1 *The power gain of any circuit can be expressed in decibels (dB). The standard of comparison of power gain in many amateur antennas is the dipole which is considered to have a gain of unity (one), or zero dB. This chart shows the relationship between power gain and decibel gain. A power gain of two, for example, is equivalent to a 3 dB gain.*

Power Gain of an antenna refers to the increase in radiated power, or received energy, in a certain direction as compared to a reference antenna (often a dipole). Gain is expressed in **decibels**(dB) and when a comparison dipole is the standard of reference, the gain figure is expressed as dBd. The relationship between decibel and power is shown in Fig. 1.

In comparing published power gains of various antennas, always be suspicious of unexpectedly high gain figures. Antenna gain, like automotive horsepower or "music power" of a stereo amplifier, is sometimes the brain child of the advertising department rather than measurements conducted in a test laboratory!

Beam Width refers to the width in degrees of the main radiation lobe of the antenna. In other words, how many degrees can you turn the beam off a given signal before the signal starts to drop in strength by a given amount? Many beam designers specify the angle from the centerline of

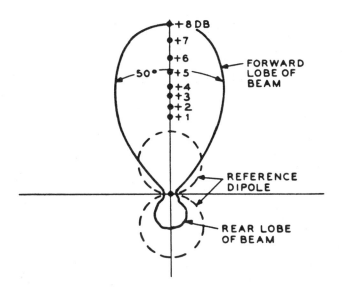

Fig.2 *The polar plot of a 4-element VHF Yagi beam. It has a half-power beamwidth of about 50 degrees at angles that represent a drop in signal strength of 3 dB. In this example, the gain of the beam is 8 dB, so the "minus 3dB" gain figure is 5 dB. The nose of the beam is quite broad making precision aiming unnecessary. The rear lobe of the beam is reduced about 12 dB below the forward lobe.*

radiation at which the signal drops 3 dB, or about 1/2 S-unit (Fig. 2). Thus, if a received signal drops 3 dB at 25 degrees each side of the centerline of maximum received signal strength, the total beam width is said to be 50 degrees.

Front-to-Back (F/B) Ratio is defined as the ratio of power radiated in the forward direction of the beam as compared to that radiated to the rear. Ratios of 12 to 30 dB (two to five S-units) are commonly observed with simple VHF beams. Front-to-back ratios measured on amateur antenna installations by the S-meter of a nearby receiver can vary widely from these figures as a result of wave reflection from the ground and nearby objects and the lack of accurate calibration of most S-meters.

Computer studies of the front-to-back ratio of a beam indicate that the maximum value is obtained at only a single frequency. While forward gain adjustments to a Yagi beam are rather uncritical, adjustments for best front-to-back ratio are quite critical. Reflections from hills and other objects can blur a good front-to-back ratio and most amateurs opt for maximum gain and do not try to adjust their beam for this elusive quality.

Frequency Span of a Yagi Beam

There are limits to the frequency range over which a beam antenna works properly. These limits are the result of the parasitic elements slowly going out of tune as the antenna operating frequency departs from the design frequency.

In general, the frequency span of a beam is plus or minus 1 to 1.25 percent of the design frequency. The total operating bandwidth, therefore, is twice these figures, or 2 to 2.5 percent. For example, a Yagi beam designed for 146.0 MHz may have an operating bandwidth of 2.5 percent centered on the design frequency. This range covers 144.175 to 147.825 MHz, or nearly the whole 2 meter band. And since the operating characteristics don't deteriorate abruptly at the bandwidth edges, this design will function well over the whole band. A drop-off in gain and front-to-back ratio will be noticed near the band edges.

Most amateurs cut their VHF beams to "where the action is". Two meter DXers cut their beams to about 144.1 MHz. And most other DX operators cut their beams to the weak signal portions of the other VHF bands.

The Change Factor for Element Lengths

Element spacing is not critical in a VHF beam but element lengths must be adjusted for each megahertz change in design frequency to achieve optimum results.

In the 50 MHz band, for every megahertz change in design frequency, all element lengths must be changed by about 2 inches, depending upon element diameter. "Fat" elements require less change than "Thin" ones. (The beams shown in this chapter can be considered to have "thin" elements.) The elements must be lengthened for a decrease in frequency and shortened for an increase in frequency.

At 144 MHz, the approximate element change factor is 1/4-inch per megahertz; at 220 MHz, 5/32-inch; and at 420 MHz, 1/16-inch.

Some useful dimensions for vhf antennas and transmission lines.

Frequency	50 MHz Inches	(cm)	146 MHz Inches	(cm)	222 MHz Inches	(cm)	432 MHz Inches	(cm)
1 wavelength [1]	236.2	(600)	80.8	(205.2)	53.2	(135.1)	27.3	(69.4)
5/8 wavelength	147.6	(375)	50.5	(128.4)	33.2	(84.4)	17.1	(43.4)
1/2 wavelength	118.1	(300)	40.4	(102.7)	26.6	(67.5)	13.7	(34.7)
3/8 wavelength	88.6	(225)	30.3	(77.0)	19.9	(50.6)	10.2	(26.0)
1/4 wavelength	59.0	(150)	20.2	(51.3)	13.3	(33.7)	6.8	(17.3)
1/2 wavelength coax [2]	77.9	(198)	26.6	(67.7)	17.5	(44.6)	9.0	(22.9)
1/4 wavelength coax [2]	38.9	(99)	13.3	(33.8)	8.7	(22.1)	4.5	(11.4)
1/2 wavelength twinlead [3]	96.8	(246)	33.1	(84.1)	21.8	(55.4)	11.2	(28.4)
1/4 wavelength twinlead [3]	48.3	(123)	16.5	(40.0)	10.9	(27.7)	5.6	(14.2)
Per cent change each MHz [4]	2		0.6		0.45		0.25	

Notes: (1) Dimensions are based on the wavelength-in-air formula 11810/F(MHz).

(2) Coaxial-cable lengths are based on the velocity factor of RG-8/U or RG-58/U, which is approximately 0.66.

(3) Twinlead lengths are based on the velocity factor of 0.82 for common TV lead-in of parallel wires with flat, solid-dielectric insulation.

(4) The dimensions given are for frequencies in the most-often used part of the bands. The dimensions *increase* as you go lower, and *decrease* as you go higher in frequency.

Table 1   Useful dimensions for VHF antennas and transmission lines.

As to the 902 and 1240 MHz bands, the change factor depends upon the element diameter and mechanical assembly of the antenna and cannot be easily predicted. It should be pointed out that a Yagi antenna functions over only a portion of the higher VHF bands (approximately 12 MHz at 902 MHz and about 18 MHz at 1240 MHz). But the antennas are so small, you can have two or three for each band, each one cut to a different design frequency.

VHF Dimensions

Some important dimensions for the VHF bands are listed in Table 1. Actual dimensions for VHF beams are derived from these fundamental figures. It is common practice to design and test an HF beam at one frequency and then scale the dimensions to another frequency. This technique can be adapted to the VHF region if element diameter and mounting hardware are scaled along with element length and spacing.

Much of the pioneering work on VHF Yagi design was done by the National Bureau of Standards, which demonstrated the interrelationship between element lengths, diameters, and spacings, as well as the effect of a metal supporting structure on antenna dimensions. The information in this section was derived from that study.

WARNING! Antennas are electrical conductors so any contact with power lines can result in death or serious injury to the installer. No antenna should be installed in a location where there is the possibility of accidental contact with utility lines or electrical service drops to buildings. Make sure the antenna, support mast, guy wires and/or tower can be erected and taken down without coming in contact with power lines. And keep metal ladders well clear of power lines!

A 3-element Yagi Beam for 50 MHz

This simple, inexpensive beam provides excellent gain and good front-to-back ratio (Fig. 3). The design frequency is 50.1 MHz and the beam is usable from 50 to 51 MHz. If it is desired to move the beam higher in frequency, each element should be reduced in length 2 inches for each megahertz change in frequency.

The boom is an 8-foot length of 1-1/4 inch aluminum tubing having a wall thickness of .058-inch, or more. Each element is made of a single length of 1/2-inch diameter aluminum tubing, with a wall thickness of .035-inch, or greater. The driven element is split in the center and insulated from the boom to provide a feedpoint.

The reflector and director mounting plates are cut from 3/16-inch aluminum stock and measure 2" x 8" in size. TV antenna assembly clamps hold the element to the plate and muffler clamps affix the plate to the boom. The muffler clamps should be plated to retard rust (Fig. 4).

The halves of the driven element are mounted on a 3/8-inch thick phenolic insulating plate by means of U-bolts. The plate is 2" x 8" in size. A wood, LEXAN, lucite or plexiglas rod that makes a slip-fit within the element segments and about 12" long is inserted into each half of the driven element to achieve rigidity. Two U-bolts, one for each element half, mount the element to the insulating plate. The wood rod is given two coats of spar varnish to make it waterproof.

The matching coil (L) is connected across the element terminals at the point where the coax line is attached. Squeezing or expanding the coil turns will drop the SWR on the coax line to a minimum at the design frequency. A slight adjustment of element length may help to achieve a low value of SWR near 50.1 MHz.

To provide proper line balance, the coax is wound into a four turn coil, about 6 inches in diameter, just before it is attached to the antenna. The coil can be taped to the antenna boom.

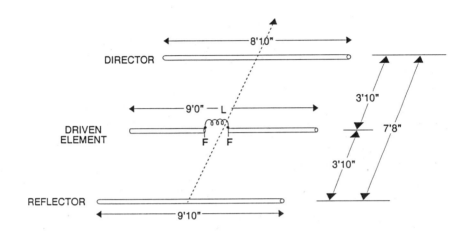

Fig.3 *A compact and effective 3-element beam for the 6 meter band. Boom is only eight feet long and the antenna may be turned with a good quality TV rotator. The matching coil L is 4 turns no. 14 solid, formvar coated wire, 1/2-inch diameter and 5/8-inch long.*

The beam is affixed to the mast by means of four U-bolts and a mounting plate which measures 6" x 3" with a 1/2-inch flange. The flange is filed to make a close fit with the mast.

For power up to 500 watts, RG-58A/U or RG-58/CU cable is adequate. RG-8A/U or RG-213/U should be used for higher power levels, or for transmission line runs over 50 feet.

For best results, the beam should be mounted in a horizontal position, at least 25 feet above ground. It can be turned with a good quality TV rotator.

A 3-element Vertical Beam for 144 MHz

Difficult to break that distant repeater? Can't get full quieting on your local repeater with a ground plane? Here's a simple 3-element vertical Yagi array that covers the whole band and provides the DX-punch to your signal! Exceptional bandwidth is achieved through the use of wider than normal element spacing.

Fig.4 *Assembly of element to boom. U-bolts fasten the element to the plate and muffler clamps hold plate to the boom. Driven element is mounted on insulating plate, as explained in text.*

The assembly is shown in Fig. 5. The boom of the array extends out beyond the reflector so that the antenna can be mounted to a mast without the mast interfering with the beam pattern.

The boom is made of 1-1/4" diameter aluminum tubing. The reflector and director elements are made of 3/32 to 1/8-inch diameter aluminum rod mounted on small insulating blocks. The driven element is made of 7/16-inch diameter aluminum tubing split at the center, as was done for the 50 MHz beam. An impedance matching "hairpin" of no. 12 enamel or Formvar coated wire is connected across the element terminals at the point the coax line is attached (Fig. 6).

If mounting blocks cannot be purchased, a satisfactory block can be made of insulating material, as shown in Fig. 7.

To provide proper line balance, the coax is wound into a four turn coil, about 4 inches in diameter, just before it is attached to the antenna boom. For power levels up to 100 watts and runs less than 30 feet, RG-58C/U coax is adequate. The beam should be mounted in the clear and as high as possible for best results.

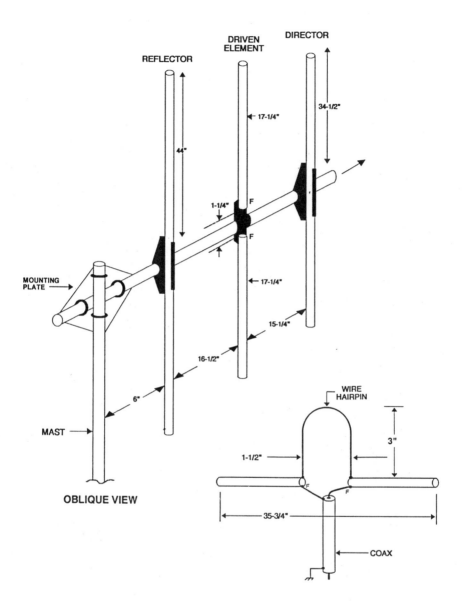

Fig.5 Compact 2 meter beam designed for vertical mounting. Boom of the array extends out behind the reflector so that the antenna can be mounted to a mast without it interfering with the beam pattern. The feedpoint matching inductor is U-shaped and is about 3 inches long and 1-1/2 inches wide.

Fig.6   *Matching inductor and SO-239 coax receptacle are mounted to bolts that attach element halves to phenolic mounting support.*

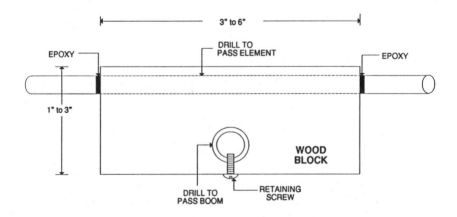

Fig.7   *Homemade mounting block for VHF antenna elements.*

A 5-element 144 or 220 MHz Yagi You Can Build

This inexpensive beam antenna is suitable for serious DX operation. It provides about 9 dB gain over a comparison dipole. That makes a 10 watt transmitter sound almost as loud as a 100 watt rig! It is designed to be mounted in the horizontal plane, but may be mounted vertically in the manner shown for the 3-element beam, if desired (Fig. 8).

The parasitic elements are made of 3/32-inch diameter aluminum or brass tubing or rod. Aluminum welding rod is cheap and satisfactory. The driven element is made of 3/8-inch tubing for the 144 MHz array and 1/4-inch tubing for the 220 MHz antenna.

A major stumbling block in home-built antennas is the problem of aligning the elements in the horizontal plane when a round boom is used. It is difficult to align the holes when the elements are passed through the boom and time-consuming to make insulating blocks to mount them atop the boom. A satisfactory answer to this problem is the use of a wood boom that can be accurately drilled on a small drill press. The round, wooden closet pole obtainable at most building supply houses makes an ideal antenna boom. When you buy one, ask the salesperson to pass the pole along a table saw to cut slices off each side to provide two parallel surfaces, as shown in the drawing. These surfaces allow you to accurately drill element mounting holes with a small drill press. Drill 3/32-inch holes along the pole through which to pass the elements, then give the pole two coats of clear spar varnish to protect it from moisture and prevent warping. The elements are cut to length and make a press-fit in the holes. A dot of epoxy cement on each element where it passes through the boom holds it securely in place.

The feed system is shown in Fig. 9. The driven element is tapped symmetrically each side of center by wires which attach to the coax line and a coax balun. This arrangement is called a **shunt-fed delta-match**. Moving the element taps back and forth along the driven element, and changing the length of the delta wires, determine the match and provide a convenient adjustment for lowest feedline SWR.

The balun consists of an electrical half-wavelength of RG-58A/U coax, coiled up and taped to the antenna boom. The outer shields of the balun ends and the transmission line are connected together. The end of one inner conductor of the balun attaches to one delta wire, and the other end is attached to the opposite delta wire and also to the inner conductor of the transmission line. This provides a good match between the balanced dipole and the unbalanced coax feed system. Care must be taken to seal the ends of the coax sections to prevent water from entering the lines.

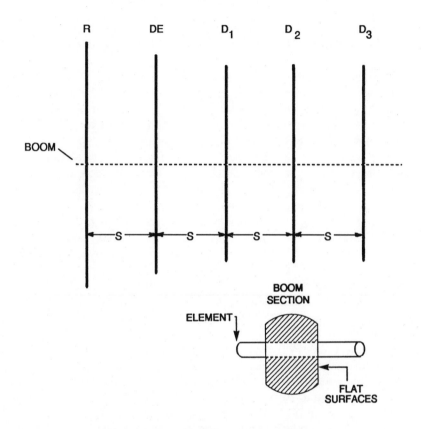

BEAM DIMENSIONS (144.1 and 220.1 MHz)
(inches)

ELEMENT	R	DE	D1	D2	D3	S
144MHz	40"	38 1/4"	36 7/8"	36 11/16"	36 7/8"	16 3/8"
220MHz	26 1/8"	25"	23 3/4"	23 11/16"	23 3/4"	10 3/4"

NOTE: D2 SHORTER THAN D1 OR D3

Fig.8 The 5-element beam for 144 or 220 MHz uses a wood boom made from a round, wood closet pole. The pole is cut to provide two flat, parallel faces to allow the element holes to be drilled accurately. Elements are cut to length and make a press-fit in the boom mounting holes.

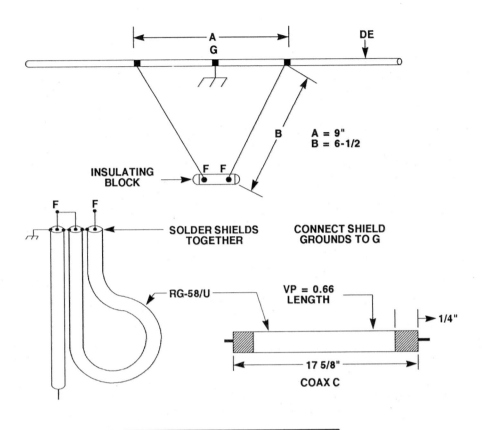

FREQ.	A	B	C
144.1	13 3/4"	10"	27"
220.1	9"	6 1/2"	17 5/8"

Fig.9 *Driven element and feed system for 144/220 MHz Yagi beam. Delta wires are attached to driven element by small clamps on each side of center point. Outer shields of feedline and balun are soldered together and then connected to midpoint of driven element (G) by means of a short wire. Approximate dimensions for delta match are given above. Coax balun (C) is electrical half-wavelength long. Balun and delta wires are connected at F-F on insulating block attached to boom.*

Adjusting the Antenna. The antenna is pre-cut for 144.1, 146.1 or 220.1 MHz. If you wish to move the design frequency higher in the band, you must trim all elements an equal amount. For the 2 meter band, trim each element length by 1/4-inch for each megahertz of upward frequency shift. For the 220 MHz band, trim each element length by 5/32-inch for each MHz of frequency shift. For example, if you wish the design frequency to be 224.1 MHz, all elements must be shortened by 4×1/8=1/2-inch.

The antenna is supported at its center of balance. If TV-clamps are used to hold it to a metal mast, a U-shaped metal plate the width of the boom and about three inches long should be bent up. It is slipped over the boom at the point where the mounting holes are drilled. This prevents crushing the boom when the clamps are tightened.

Before you adjust antenna elements, run an SWR check across the band to determine the actual resonant frequency of the array.

The SWR Check

The following test should be run with an SWR meter rated for operation up to 150 MHz for the 2 meter antenna, or up to 250 MHz for the 220 MHz antenna. The SWR meter is placed in the coax line to the antenna. An SWR measurement is made every 500 kHz across the band and the SWR measurements plotted on graph paper. A smoothed curve is drawn through the measured points. The point on the curve of lowest SWR is the design frequency of the array. For the 2-meter antenna, this frequency should be near 144.1 MHz (Fig. 10).

If the SWR is too high at this point, it may be reduced by altering the tap points of the delta wires. Larger changes can be made by varying the delta wire length. Once the point of minimum SWR has been established, the resonant frequency can be moved about by changing the length of the driven element slightly. For changes greater than 1 MHz, it will be necessary to alter the length of all elements, as discussed earlier.

A 12-element "Big Gun" for 220/223 MHz

This long Yagi array will make a 10 watt signal sound like a 200 watt block buster! In addition, it will boost received signals 4 to 5 S-units over a simple ground plane antenna. It is easy to build and very inexpensive (Fig. 11). The beam consists of ten directors, a driven element, and a reflector; all mounted on a 10-foot wood boom. Dimensions for the low and high ends of the band are given in the illustration.

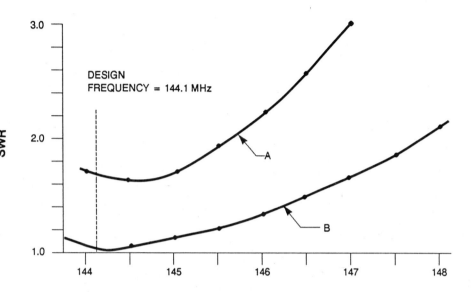

Fig.10   *SWR curves run on delta-matched beam. The first curve showed the SWR is too high at the design frequency of 144.1 MHz. Sliding the element taps about 3/4-inch in towards the center of the element resulted in the second curve. Notice that the frequency of lowest SWR was changed slightly by this adjustment.*

Dimensions for the shunt-fed delta match are given in the drawing. Adjustments, if any, are carried out in the sequence described for the previous antennas. Balun dimensions are shown in the drawing. Care must be taken to seal the ends of the coax sections so that water does not enter the lines.

Element lengths are critical and are given to the nearest sixteenth-inch. The parasitic elements are made of aluminum welding rod (3/32 to 1/8-inch diameter) and the driven element is made of 1/4-inch diameter aluminum tubing. The elements are cut slightly longer than specified and are trimmed to length with a file. The ends should be square, with a very slight rounding of the edge done with a fine file. It is a good idea to number or color code each element with a spot of paint as it is very difficult to tell one from another when they are completed!

The boom is made from a dry, well-seasoned length of wood. It should be straight, with no knots. Alternatively, some builders cut up used fiberglass

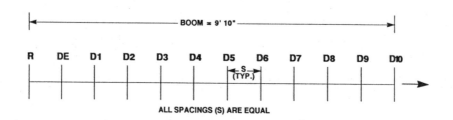

BEAM DIMENSIONS IN INCHES

f (MHz)	R	DE	D1	D2	D3, D10	D4, D9	D5 to D8	S
220.1	26 1/16"	25"	24 1/16"	23 5/16"	22 15/16"	22 1/2"	22 1/8"	10 3/4"
223.1	25 7/16"	24 1/2"	23 9/16"	22 13/16"	22 7/16"	22"	21 9/16"	10 1/4"

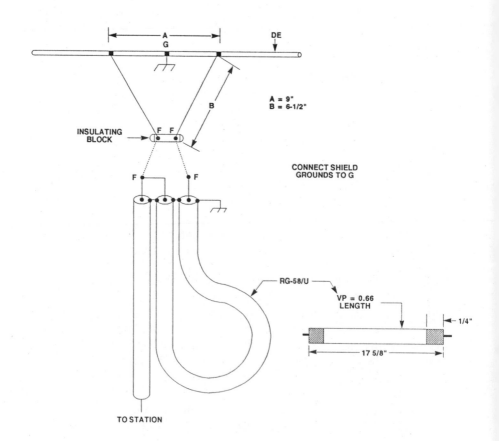

Fig.11 *Element dimensions and matching system for 220 MHz, 12 element Yagi beam. Shields of coax line and balun are connected to midpoint of driven element (G) by short wire. Tap points of delta wires on drive element are adjusted for lowest SWR on feedline at design frequency of the antenna.*

vaulting poles for antenna booms. If a wood boom is used, apply two coats of spar varnish after drilling to protect it from the weather.

The elements are held in position with a drop of epoxy cement. Care should be taken in drilling the boom holes so that the elements lie in the same plane and element spacing is correct.

The shunt-fed delta match, described for the 5-element beam, is used in this array. The approximate dimensions are given in the drawing. Placement of the taps on the driven element (the spread) and the length of the delta wires determine the match and provide a convenient means of adjustment for lowest feedline SWR. A small block of lucite, plexiglass or LEXAN is fastened to the boom by means of wood screws. Bolts mounted in the block hold the delta wires to the balun and feedline.

The antenna is supported at the center of balance. The balun and RG-213/U (or RG-8A/U) feedline are taped to the boom and run forward to the mast. The line is then brought down the mast to the station. If mounting is vertical, the feedline should be brought out the rear of the array as shown in chapter 7. The top two feet of the supporting mast should be made of varnished or painted wood so that no metal is near the elements of the antenna.

Adjusting the antenna. The station transmitter is set at the design frequency (220.1 MHz, for example) and the antenna is fed a few watts of power through a VHF-type SWR meter. If the SWR is less than 1.5-to-1, no further adjustment is necessary.

Adjustment of the driven element tap points will lower the SWR, if necessary. The adjustments are made equally to each tap point so that the taps remain equal distances from the center of the driven element. Changing the length of the delta wires should be done only if tap point adjustments fail to provide a satisfactory value of SWR.

The adjustments are made with the antenna mounted atop a wood stepladder. The antenna is pointed up at the sky at a 45-degree angle for ease of adjustment and to prevent interaction with nearby metallic objects.

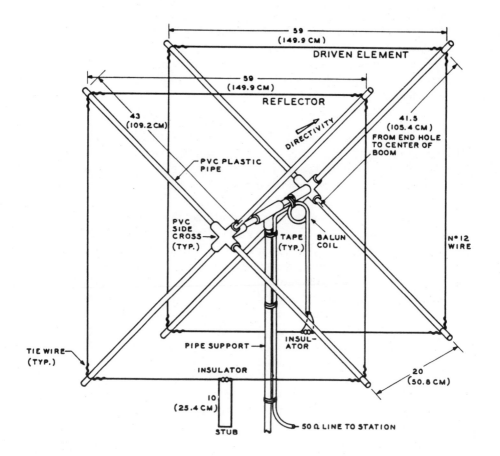

Fig.12 A simple 2 element Quad for 6 meters. The framework is made of 1/2-inch plastic PVC water pipe with length of one inch PVC pipe for the boom. The Quad loops are strung through holes drilled in ends of arms and wires are under enough tension to bow arms slightly inward. A stub with a moveable shorting bar tunes the reflector element. Quad may be tuned for maximum forward gain on a received signal. After optimum stub length is found, the bar is soldered in position.

The VHF Quad Antenna

The popular **Cubical Quad** antenna is composed of two or more spaced loops of wire, each approximately a quarter wavelength on a side. Multi-loop Quads have proven effective on the high frequency bands, and many are in use on the 6 and 2 meter bands. Quad antennas may be horizontally or vertically polarized by feeding the driven loop in a horizontal or vertical leg, and the loops may be square, triangular or diamond shaped.

The Quad antenna has the same, or slightly more, gain than a Yagi with the same number of elements. It also has greater bandwidth (lower SWR over a given frequency range) than the Yagi, requires little or no adjustment, and requires no aluminum tubing, which is difficult to obtain in some areas.

A Quad Beam for 50 MHz

The simplicity of construction and low cost of the Quad makes it an attractive antenna project for the homebrew artist. This is a good, all-around, versatile antenna for the beginner.

A design for the 6 meter band is shown in Fig. 12. The arms of the Quad framework should be made of nonconductive material, such as bamboo or plastic pipe. The framework described here uses 1-inch PVC (polyvinyl chloride) pipe for the boom and 1/2-inch PVC pipe for the crossarms. Plastic pipe "sidecrosses" fasten the arms to each other and to the boom with the aid of epoxy cement. The boom is mounted to the mast with a T-fitting, or with a metal plate and U-bolts.

The PVC crossarms are reinforced at the ends with 12-inch lengths of wood dowel rod placed inside the tubes and epoxied in place. The Quad loops are strung with enough tension to bow the arms slightly. Each loop is made of no. 14 solid enamel covered wire and is terminated in a small insulator. A 10-inch wire stub with a movable shorting bar tunes the reflector loop.

The Quad in the drawing is horizontally polarized. All that is necessary for vertical polarization is to rotate the Quad a quarter-turn so the feedline and stub are on one side of the assembly instead of the bottom.

Placement of the shorting bar on the stub is not critical. It may be soldered in place about eight inches from the insulator. Or, it may be adjusted for maximum received signal while listening to a local station. After optimum placement is found, the stub is soldered in place. When checked with a dip oscillator, the driven element should be resonant near 50.5 MHz

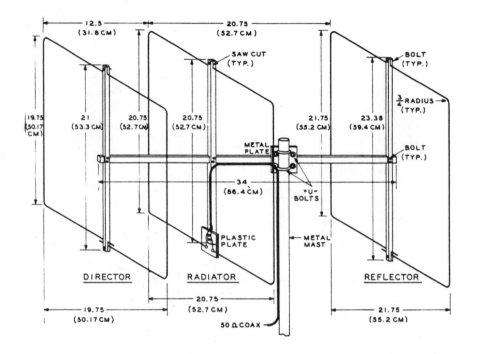

Fig.13 Three element Quad for 2 meters. This inexpensive design provides horizontally polarized signals when fed at the bottom of the radiator loop. The antenna may be rotated 90 degrees along the boom axis to provide vertical polarization. For final adjustment, the boom is slid back and forth slightly in the mounting clamp for minimum SWR reading on coax line.

(with feedline disconnected and a short across the feedpoint). Reflector loop resonance falls at about 48 MHz when the stub is properly adjusted.

The Quad is fed with 50 ohm coax line which is coiled into a 5-turn balun choke coil about 5 inches in diameter. Placement of the balun is shown in the illustration.

A 3-element Quad for 2 Meters

Shown in Fig. 13 is a Quad designed for the 144 MHz band. It is built of inexpensive aluminum clothesline wire and wood. The assembly is shown horizontally polarized, but rotating the array a quarter-turn about its axis will provide vertical polarization. Antenna gain is about equal to a 4 element Yagi beam.

The Quad boom is a length of smooth, finished "one-by-one" lumber (which measures about 7/8-inch square) as are the element supports, which are tapered to 1/2-inch at the tips. All wood parts are given several coats of spar varnish for weather protection.

Each element support is notched at the point it is fastened to the boom. The notches in the reflector and director supports are centered, but the driven element notch is 11-inches from the top. A slight step cut in the bottom of the support accomodates the insulating plate that supports the ends of the driven loop.

Holes drilled near the top of the supports, and bisected lengthwise by a saw cut, accomodate the aluminum element wires. At the bottom of the parasitic element supports two small holes hold the ends of the loops. Small bolts and nuts pinch the saw slots together to anchor the loops in place.

The aluminum wire is cut slightly longer than needed and the segments are bent into rectangles having the dimensions shown. Crimp   terminal lugs on the ends of the driven loop and bolt the lugs to screws on the terminal block. Push the ends of the parasitic loops through the holes in their supports. The adjacent ends of these loops are connected together with a short aluminum strap held in place with 4-40 nuts and bolts.

The boom of the Quad is fastened to the support mast about 8 inches behind the driven loop. For final adjustment of the SWR on the coax feedline, the boom is moved back and forth a few inches in the mounting clamp, changing the distance from the metal mast to the driven element. Do not substitute wood for the metal mast as the loop dimensions are adjusted to compensate for the mast.

The RG-58A/U coax line is coiled into a 4-turn balun choke, about 4 inches in diameter, just before it attaches to the driven loop element.

Antenna Construction

Building VHF antennas is fun, and much of the construction material is available from TV and electronic supply houses in the form of standard TV antenna hardware. Local hardware and hobby shops, as well as surplus metal outlets, are good sources of aluminum tubing and rod. Thus, unlike some of the more exotic VHF components, material for VHF antennas is relatively easy to obtain.

The mark of a good VHF antenna is that it will stay up, even in a windstorm, and that it is easy to take down when you are finished with it. The secret to success is the construction technique you use.

Starting with the mast, Fig. 14 shows three methods of joining the antenna array to the support structure. The double gusset plates are recommended for a large array or a windy location. A variation on this design is to have a single plate and affix it between the mast and boom with U-bolts. All mounting hardware should be plated or otherwise finished to protect against rust and corrosion.

The various elements of a typical VHF Yagi beam are generally made of smaller diameter material than the boom and present a mounting and support problem. Fig. 15 shows four methods of mounting an element to a round, metal boom.

In illustration A the element is passed through the boom. This is practical if the boom diameter is at least four times the element diameter. The presence of the metal boom requires that the element be lengthened by an amount equal to 0.7 times the boom diameter.

If it is desired to insulate the element from the boom, it can be mounted above the boom on an insulating plate. The element is held to the plate with small aluminum straps formed around the element and bolted to the plate, as shown in illustration B.

In drawing C, the element is held to the boom with U-bolts passed through the element, with stiffening half-diameter tubing sections placed above and below the element. A saddle bracket holds the element firmly to the boom.

A combination U-bolt and clamp holds the element to the boom, as shown in drawing D. The U-bolt and saddle bracket mate with straps cut from aluminum tubing that encircle the element.

METHODS FOR JOINING MAST AND SUPPORT SECTIONS

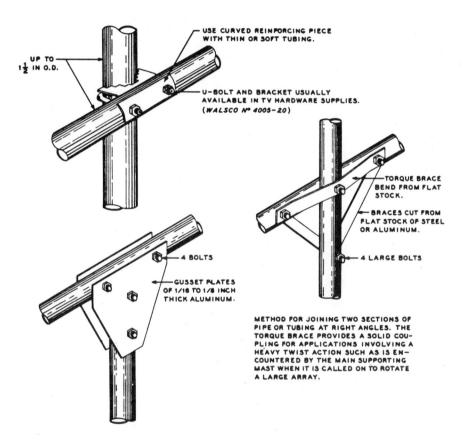

USE CURVED REINFORCING PIECE WITH THIN OR SOFT TUBING.

UP TO $1\frac{1}{2}$ IN O.D.

U-BOLT AND BRACKET USUALLY AVAILABLE IN TV HARDWARE SUPPLIES. (WALSCO Nº 4005-20)

TORQUE BRACE BEND FROM FLAT STOCK.

BRACES CUT FROM FLAT STOCK OF STEEL OR ALUMINUM.

4 BOLTS

GUSSET PLATES OF 1/16 TO 1/8 INCH THICK ALUMINUM.

4 LARGE BOLTS

METHOD FOR JOINING TWO SECTIONS OF PIPE OR TUBING AT RIGHT ANGLES. THE TORQUE BRACE PROVIDES A SOLID COUPLING FOR APPLICATIONS INVOLVING A HEAVY TWIST ACTION SUCH AS IS ENCOUNTERED BY THE MAIN SUPPORTING MAST WHEN IT IS CALLED ON TO ROTATE A LARGE ARRAY.

ARRANGEMENT WHICH MAKES AN EFFECTIVE T-JOINT WHEN TUBING IS OF SAME DIAMETER.

Fig.14 *Mechanical assembly of antenna boom to mast is important if array is to withstand bad weather. Brackets or gusset plates provide rigid, rugged support where needed.*

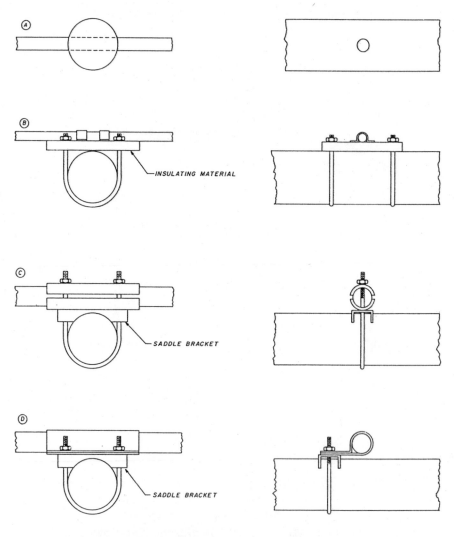

Fig.15   *Typical element mounting methods:*
a) *Element passed through the boom, insulated or uninsulated.*
b) *Element insulated and mounted above the boom.*
c *Element held in contact with the boom by U-bolt, saddle bracket and clamps.*
d) *Element mounted by bracket and clamp.*

Chapter 9

VHF INTERFERENCE AND HOW YOU CAN SUPPRESS IT

The VHF operator sometimes discovers to his chagrin that his transmitter is causing television interference (TVI), stereo interference or, even worse, telephone interference!

The causes of VHF interference are similar those experienced on the HF bands. 1)- The signal of the amateur transmitter contains harmonics or spurious signals that show up as a crosshatch pattern or modulation on nearby entertainment equipment. 2)- The signal of the amateur transmitter is so strong that it overloads (blocks) the nearby TV receiver, causing crosshatching or a complete blackout of the picture. 3)- The signal of the amateur transmitter is picked up by the entertainment equipment due to poor shielding or poor image response of the television set. For example, one common type of image interference shows up on TV channel 2 when 2 meter signals are received as an image on a TV set with a 42 MHz intermediate frequency.

Suitable remedies can be applied to the transmitting equipment to lessen the probability of harmonic or spurious signal interference and a reduction of radiated power usually reduces TV receiver overload problems. Image, stereo and telephone interference, however, are problems of the affected equipment and must be dealt with on a case-by-case basis. This is a public relations job as well as a technical problem, and requires a great deal of tact and patience on the part of the amateur. The TV viewer or stereo buff usually doesn't have the faintest concept of amateur radio, or the complex problem of interference between nearby transmitters and poorly designed entertainment equipment. As long as **your** signal interferes with **his** equipment, he automatically assumes the fault is yours! After all, didn't he pay a bundle of bucks for his TV or stereo?

Unfortunately, when entertainment equipment is built for a highly competitive market, cost cutting is rampant. One of the first cost reductions taken is to eliminate shielding and filtering that could protect the equipment from interference. This situation will continue until Federal legislation requires mandatory shielding and filtering on home entertainment equipment to reduce its susceptibility to strong off-frequency signals.

Generally speaking, the VHF operator running 100 watts or less and operating in a medium-to-strong TV signal area should experience few TVI problems. Exceptions to this statement exist, particularly on 6 meters. TV receiver overload, weak TV signals in a fringe area, excessive harmonics, or a defective TV antenna installation can damage reception even when the amateur is running low power.

It can be said that TVI/RFI is no problem if you don't have it, and HELL if you do! The following sections provide TVI suppression information, with the author's sincere hope you will never need it!

Harmonics

All radio transmitters generate **harmonics** to a greater or lesser degree. A harmonic signal is a multiple of the transmitted signal. For example, if your transmitter is on 50 MHz, it generates small signals on 100, 150, 200 MHz, and so on. Most harmonics are quite weak, but some are not. The strongest are the second (100 MHz) and third (150 MHz). But all harmonics can be troublesome, particularly if they fall on a frequency used by another communication service. On a TV set, the harmonic shows up as a crosshatch pattern, changing in appearance when the transmitter frequency is shifted.

Harmonic Relationships

A chart of the various VHF TV and FM channels is shown in Fig. 1. As an example, the second harmonic of a 50 MHz (six meter) station falls at 50 x 2 = 100 MHz, which is in the middle of the FM broadcast band. This could be a source of interference. In addition, the fourth harmonic of a six meter station (50 x 4 = 200 MHz) falls within the range of TV channels 11 to 13.

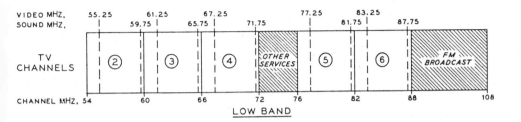

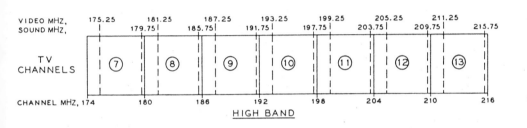

Fig.1 *VHF FM-TV chart showing video and sound frequencies for each channel. The video signal is at the low frequency side of the channel and the audio signal is at the high frequency side. The second harmonic of a 50 MHz transmitter falls in the FM broadcast band near 100 MHz and the third harmonic falls within the range of TV channels 11 ,12, and 13. The 50 MHz transmitter is adjacent to the video signal of TV channel 2. Channel 13 overload can be caused by nearby 220 MHz transmitter.*

TV Receiver Blocking

A potent cause of conflict between amateurs and nearby TV viewers is the fact that many TV sets do not have sufficient front-end selectivity to reject strong signals outside the TV channels. A case in point is the 6 meter amateur band which is in close frequency proximity to TV channel 2. If the TV receiver is overloaded by a nearby 6 meter transmitter, the interference may vary from a light cross-hatching to a complete blackout of the picture. Audio interference may accompany the picture trouble. In severe cases, channels 3 through 6 may also be affected. Blockng of channel 13 by a 220 MHz transmitter is similar in nature, but usually not so severe.

Thus, the 6 meter operator has three areas in which he might find himself in trouble: in the FM band, in channel 2, and in channels 11 and 12. In addition, his 6 meter signal may be getting into poorly shielded stereo systems or the neighbor's telephone!

The 2 meter operator is in better shape. His second harmonic (144 $\times 2 = 228$ MHz) does not fall in the VHF TV channels, nor is the 2 meter band adjacent to any TV channels. He still might encounter stereo or telephone interference problems. And his fourth harmonic falls in UHF TV channels 31-34.

So far, interference has not been a major problem on the higher VHF bands, possibly because most operators run low power. But the third harmonic of a 220 MHz transmitter falls in UHF TV channels 45-48 and the second harmonic of a 420 MHz transmitter falls in UHF TV channels 75-80.

Coax and Stripline Filters

An effective VHF filter can be built that prevents harmonics in the transmitter from reaching the antenna. A filter is also helpful in receiving, as it rejects out-of-band signals that might otherwise overload the receiver.

An easy filter to build is the coaxial design, which is a linear tuned circuit an electrical quarter- or half-wavelength long (Fig. 2). These filters pass only a very narrow band of frequencies and reject all energy outside the passband. The quarter-wavelength filter (A) is a resonant line, grounded at one end, and inclosed in a shield. The half-wavelength filter (B) is grounded at both ends of the resonant line. Both filters are tuned by means of a variable capacitor placed at the high impedance point of the resonant line. Energy is coupled into and out of the filter by coupling loops placed at the low impedance points of the line.

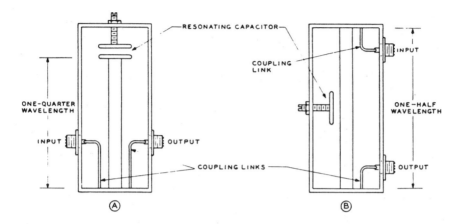

Fig.2 *The coaxial filter is a tuned circuit having very high selectivity. Quarter-wavelength filter (A) consists of line tuned at end by a variable capacitor. The half-wavelength filter (B) is tuned at the midpoint. Filter is tuned to the operating frequency of the transmitter. The filter is intro-duced into the coax feedline by small coupling loops placed near the bottom (ground) end of the coaxial line.*

Filter tuning is critical, and the coupling loops must be adjusted to provide the proper degree of coupling so the filter is not loaded too heavily by the external circuits.

The **strip line** filter substitutes a flat strap for the tubular center conductor which may, in fact, be plated on a printed circuit board with the shield soldered to the ground area of the board. This construction is common in the UHF region.

A Six Meter Stripline Transmitter Filter

Shown in Fig. 3 is a transmitter filter for 6 meter operators. The filter rejects harmonics as it is a highly selective circuit tuned to the transmitter operating frequency. The filter line is folded to conserve space.

The unit is built in an aluminum box measuring 17 x 6 x 3 inches, subdivided with an aluminum partition bolted to the side and one end of the chassis. The filter line is a 1-inch wide strap of 1/16-inch thick aluminum

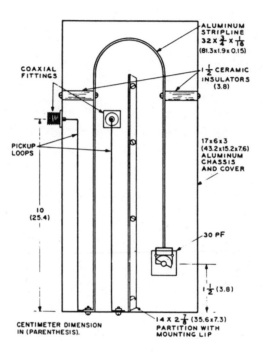

Fig. 3 *Six meter stripline version of the coaxial filter is folded to conserve space. The center conductor is an aluminum strap. Pickup loops are no.12 wire spaced 1/4-inch from the strap. Loops and strap are bolted to bottom of aluminum box. The filter is adjusted to the operating frequency by the 30 pF air capacitor at end of the line. (Dimensions in parenthesis are in centimeters).*

32 inches long. The line is supported at the bottom end by a bracket bolted to the chassis and at the opposite end by the stator of the tuning capacitor. Two ceramic insulators further support the line.

The input and output coax receptacles are mounted 10 inches from the grounded end of the line on either side of it. The pickup wires are spaced about 1/4-inch from the line.

Tuning the filter. To tune the filter, place it in the coax line between the transmitter and the antenna, with an SWR meter in the line between

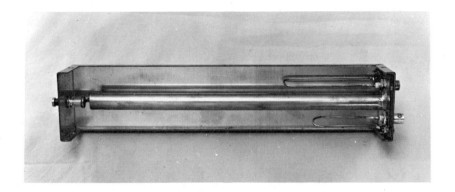

Fig.4 *A coaxial line filter for 2 meters. The copper line is soldered to a copper grounding plate (right). The pickup loops and coax fittings are also mounted to this plate, which is bolted to the end of the aluminum box. The line is tuned by a 15 pF piston-type variable capacitor at left. The coupling loops are between 3 and 4 inches long (not critical) and are parallel to the center conductor and about 1/2-inch away.*

transmitter and filter. Using low power, adjust the filter variable capacitor for lowest SWR at your chosen operating frequency. Assuming your antenna is fairly well matched and the spacing between the pickup wires and the filter line is as described, this is all there is to be done. Adjusting the wires slightly will vary the SWR to a degree. Moving the wires closer to the filter line makes filter tuning broader, but degrades filter selectivity somewhat. Conversely, moving the wires away from the line increases selectivity. When completed, a plate should be bolted to the open side of the box to complete the shielding.

A 2 Meter Stripline Transmitter Filter

The 2 meter filter is shown in Fig. 4. It uses a 12 x 2-1/4 x 2-1/2 inch aluminum box and an 11-inch length of 5/8-inch diameter copper tubing as the resonant line.

The copper line, coax connectors and grounding lugs for the two "hairpin" loops are all soldered to a square copper plate cut to fit the end of the box. The piston-capacitor may be replaced with a single bearing,

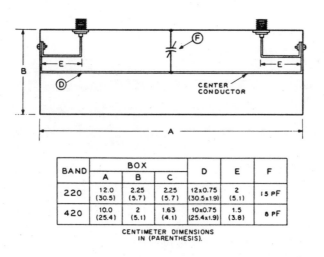

BAND	BOX			D	E	F
	A	B	C			
220	12.0 (30.5)	2.25 (5.7)	2.25 (5.7)	12x0.75 (30.5x1.9)	2 (5.1)	15 PF
420	10.0 (25.4)	2 (5.1)	1.63 (4.1)	10x0.75 (25.4x1.9)	1.5 (3.8)	8 PF

CENTIMETER DIMENSIONS
IN (PARENTHESIS).

Fig.5   Filters for 220 and 420 MHz are half-wave design. The 220 MHz model is tuned with a variable air capacitor, while a disc or piston-type capacitor is used for the 420 MHz filter. Dimension C is the width of the filter box. The coupling loops are spaced about 1/4-inch away from the center stripline.

15 pF double spaced midget variable air capacitor, if desired. The rod may have to be shortened slightly to accept a different style capacitor. The coupling loops are between 3 and 4 inches long (not critical). They are placed parallel to the center conductor and about one-quarter inch away.

Stripline Transmitter Filters for 220 and 432 MHz

The passband filter design shown in Fig. 2B is for use in the 220 and 432 MHz bands. It rejects transmitter harmonics which fall in the UHF TV bands. This is a half-wave stripline circuit, tuned with a capacitor placed at the center of the line. The filters are built in an aluminum box with the coax fittings and tuning capacitor mounted on one side of the box (Fig. 5). The center conductor is made of aluminum strap. The pickup wires (E) are spaced about 1/4-inch away from the line. In the case of the 220 MHz filter, the capacitor (F) is a variable air capacitor, while a piston-

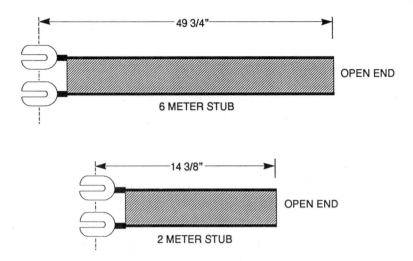

Fig.6 *Simple "ribbon" filters made of 300 ohm line are effective in reducing TV interference from nearby transmitters. Length is measured from center of lug terminal to end of line. Stubs may be used in parallel if interference is experienced from both 6 and 2 meters.*

type capacitor is used in the 432 MHz model. The capacitor is adjusted for minimum SWR on the line, as described in the previous section. Power capacity of the filter is about 100 watts. The filters are tuned in the manner described for the 6 meter filter.

TV Overload Filters for 6 and 2 meters

"Ribbon" filters are quite effective in reducing TV overload from nearby 6 and 2 meter transmitters. In the case of a TV receiver having 300 ohm ribbon feedline, the filter is a tuned length of line placed across the receiver antenna terminals (Fig. 6). One end of the filter has a pair of lugs placed on it, while the opposite end of the line is left open. The line is cut to length as shown. For best resistance to overload, line length is critical, especially in the case of the 2 meter model. You can cut the filter line about 1/2-inch longer than specified, and then trim it, 1/8-inch at a time, for best overload protection.

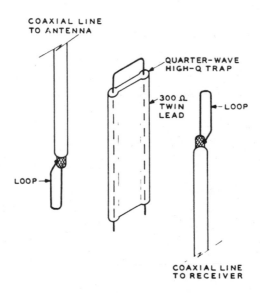

Fig.7 *A "ribbon" filter for TV coax line is made of 300 ohm line. One end of line is shorted. Coax line is cut near receiver and line ends are connected to small loops 2 inches long. Loops are taped to the ribbon filter after positions are adjusted for minimum interference.*

A simple linear filter for TV coax lead-in is shown in Fig. 7. The filter is made of ribbon line cut to length for the transmitter frequency. One end of the line is shorted, and the opposite end left open. For 6 meter overload interference, the stub is 49-3/4 inches long and for 144 MHz, stub length is 14-3/8 inches. The stubs should be made slightly longer than specified and trimmed for minimum picture interference.

The coax line to the TV receiver is cut, and the line ends are terminated in wire loops 2 inches long and as wide as the ribbon stub. The loops are placed adjacent to the shorted end of the ribbon, one on each side, and held in position with tape. The position of the loops is adjusted for minimum interfering signal.

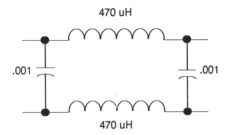

Fig.8 *Simple telephone line filter is made of miniature RF chokes and bypass capacitors. The chokes are 470 microhenries (Miller 9220-12, or equivalent).*

Telephone Interference

Telephone lines can act as an antenna, picking up your signal and conducting it into the telephone or answering machine. A telephone line filter (Fig. 8) will help to cure this problem. The filter is installed in series with the plus (green) and minus (red) balanced telephone line. Some phone systems use blue for the negative wire and blue/blue-white for the positive one. The filter is placed in the line at the instrument. In cases of severe interference, it may be also necessary to place a small .001 uF disc capacitor across the microphone and headphone terminals of the handset.

In addition to this filter, a telephone answering machine or a cordless telephone often require the power lead to be filtered. The power cord is wrapped around a 7.5-inch long, 1/2-inch diameter ferrite rod with a permeability of 800 (Amidon R-33-050-750, or equivalent). The power cord is secured at the start and finish of the single layer, close-spaced winding with cable ties.

Stereo Interference

In the majority of stereo interference cases, corrective action must be taken at the equipment experiencing the problem. Long speaker leads and wires from tuners to amplifiers act as antennas, picking up the unwanted signal and feeding it into sensitive circuits. Signals can also enter via the power cord. Once inside the equipment, the offending signal is difficult to eliminate. The addition of interference suppression circuits in the equipment itself should only be done by a qualified service technician who

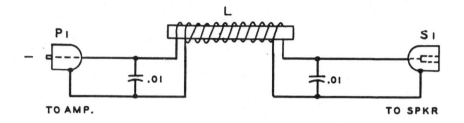

Fig.9   Lead filter for speaker wires. Capacitor is .01 uF disc ceramic. Coils are 30 to 40 turns each, wound on a ferrite rod, 0.37-inch diameter, 4 inches long, permeability 800. (Amidon R-33-037-400, or equivalent). Tape windings in place. Wire size should be no.16 or larger, depending upon amplifier power.

has the maintenance and repair manuals plus the expertise in solving RFI-induced problems.

The first step in investigating interference is to reverse the line plug of the stereo amplifier in the electric outlet. Some stereo gear has the chassis (ground) bypassed to one side of the power line and the polarity of the plug might inadvertently place the chassis above ground as far as interference is concerned. This is an easy test and it may work for you. If it does not, the next step is to temporarily wrap the line cord around a ferrite rod, as described for the telephone answering machine in the previous section. Leave this choke in place as you examine other RFI-prone circuits.

Audio and Speaker Leads. Once the power line is filtered, turn your attention to the other leads going to the stereo amplifier. First, disconnect all input leads (phono, tuner, tape deck, etc.). If interference is not reduced, the signal is probably being picked up by the speaker leads. Shown in Fig. 9 is a speaker lead filter that can be easily made in the home workshop. One filter is placed in each lead. The filter is made of a ferrite rod and matching input and output plugs that fit the stereo connectors. Two lengths of hookup wire are wrapped around the rod and taped or epoxied in place. The lead from the filter to the amplifier (P1) should be very short. The speaker leads to plug S1 can be any length.

With a filter of this type in each speaker cable, the stereo should be checked for interference before any input leads are attached.

In some cases, it may be necessary to remove the .01 uF capacitor connected across plug P1 to prevent amplifier instability.

Chapter 10

VHF ROUNDUP

VHF enthusiasts operating FM in an area of high activity can get by with almost no test equipment. Help is only a telephone call or repeater contact away. As long as the local repeater can be activated, the VHFer knows his rig is working, and when the repeater does not respond to him, he can return his transceiver to the distributor for repair or replacement. (How well the equipment is working may be another story!)

VHF experimenters, on the other hand, must depend largely upon test equipment to judge the progress of their work and to check out their gear. As a result, they are usually well-equipped with VHF test devices. Many major pieces of general test equipment that function up to 450 MHz can be bought, but specialized VHF test equipment that will function up to 1300 MHz is expensive and hard to come by. Even the most knowledgeable experimenters often return their 1296 MHz gear to the factory rather than try to repair or modify it themselves!

The SWR Meter

The electrical "pattern" formed in the coax line by the combination of forward and reflected waves provides an ideal means of measuring the degree of match between the antenna and the transmission line. The standing-wave-ratio is measured by an SWR meter (sometimes called a **directional coupler**).

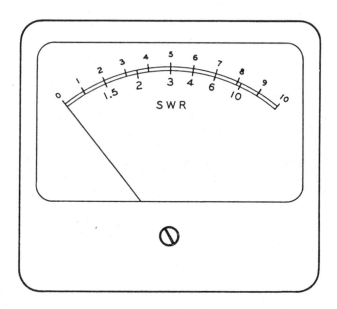

Fig.1 *Typical SWR meter calibration. The top scale is merely for reference. The bottom scale is calibrated in standing wave ratio (SWR). Lowest SWR reading is at left of scale (unity SWR). SWR readings above 10 are generally inaccurate and most meters are not calibrated above that value. The SWR meter measures forward and reverse voltage in a transmission line, but scale is calibrated in terms of SWR. Be sure the SWR meter you buy is rated for VHF operation.*

    The SWR meter provides a reading which expresses the ratio of voltages in the two waves. The SWR is read directly from a meter scale with a zero reading indicating the reflected wave is zero (SWR = unity, or 1-to-1). A full-scale reflected reading indicates a state of maximum wave reflection and a very high SWR (Fig. 1).

    With a matched, resonant antenna, the SWR should be 1-to-1. As the match between an antenna and the coax line becomes worse, the reflected wave rises in amplitude and the "reverse" meter reading increases. Generally speaking, the reverse reading for your antenna should indicate an SWR of less than 2-to-1, as most solid state transmitters are designed to work

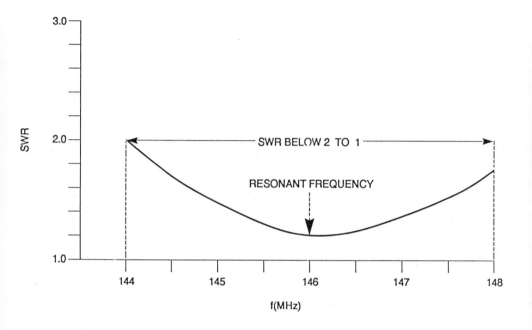

Fig.2 *The SWR reading made at intervals across the band can be plotted on a chart. The horizontal axis is frequency and the vertical axis is SWR. When the plotted points are connected by a smooth line, the result is an SWR curve for the antenna in use. The point of lowest SWR is the design frequency of the antenna. Note how the SWR rises when the transmitter is operated off the design frequency. Most VHF transmitters are rated for operation with an SWR figure of less than 2-to-1. The frequency span of the antenna which keeps the SWR below this figure is shown on the graph.*

The "Thruline" Model 43 wattmeter (Bird Electronics Corp., Cleveland OH). This instrument is suitable for use over the HF, VHF and UHF ranges. Plug-in detector heads determine operating range and power level. Scale readings are converted to SWR by means of a chart. This instrument provides a more accurate reading than most small, imported SWR meters.

with SWR values less than this, and some automatically decrease power output as SWR rises.

A big advantage of the SWR meter is that it may be left in the line as a continual monitor of antenna performance. No adjustments at the station will alter the SWR value, but changes in the antenna itself will change the reading. An abrupt change in SWR may mean a problem with the antenna.

What Does the SWR Meter Tell You?

The lower the value of SWR at the design frequency of the antenna, the better the match to the transmission line. And, the lower the value of SWR will be across the band. A representative SWR curve for a 2 meter antenna is shown in Fig. 2. The design frequency is 146 MHz and the SWR

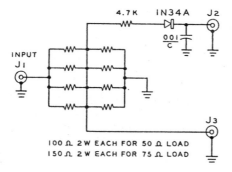

Fig.3 *Inexpensive 15 watt VHF dummy load is made up of series-parallel connected composition resistors. Power source is connected to input receptacle, J1. The relative power level can be monitored with low-range milliammeter at J2. The unit may be inserted in a coax line (J1 in, J2 out) to serve as an attenuator.*

is 1.2-to-1. Note how the SWR rises when the antenna is operated off the design frequency. The frequency span of the antenna which will hold the SWR below 2-to-1 is shown on the graph. But remember—a long run of coax can mask an antenna mismatch due to reflected power losses in the coax.

Good and Bad SWR Meters

Almost any SWR meter will work OK at the lower frequencies but only the good ones will give you meaningful readings at frequencies above 50 MHz. There are cheap and inaccurate SWR meters on the market for the CB operators. Some time ago, the author of this handbook had the opportunity to connect three CB-type SWR meters in series in one transmission line. None of the readings agreed and the variation in readings changed with the power level of the transmitter!

You can build a good VHF SWR meter or you can buy one. Several brands of SWR meters have plug-in pickup heads and are especially useful as the heads can be changed to match the operating frequency and the transmitter power level.

Fig.4 *The interior of the dummy load. Coax input recep-*
tacle is at the left, with receptacles J2 and J3 at the right.
The 2-watt resistors are mounted between thin copper
plates. The center plate has a mounting foot which is bolted
to a small ceramic insulator to provide support for the
assembly.

A VHF Dummy Load and Attenuator

The 15 watt power capability of this compact, inexpensive dummy load makes it popular for use with many of the low power VHF transceivers on the market. The load circuit is shown in Fig. 3. Two sets of four 100 ohm, 2 watt composition resistors are connected in series-parallel to form the load. Receptacle J1 is the input connection. A second receptacle (J3) permits using the device as a power attenuator. The other receptacle (J2) connects to a diode detector used to monitor the power level with a low range dc milliammeter.

Load Construction. The dummy load is built in a small aluminum box measuring about 3-1/2 x 2 x 1-1/4 inches in size. Layout of the interior can be seen in Fig. 4. Three pieces of thin, shim copper sheet join the resistors and serve as heat sinks. The outer pieces measure about 2 x 1 inches and the center piece is about 1/2-inch wider to provide a mounting foot. The foot is bolted to a small insulator to provide support for the assembly. A #60 drill is used to make the resistor mounting holes. The resistor leads are cleaned and tinned before any soldering begins. Solder #14 tinned copper wire to make connections from the mounting plates to the various receptacles.

When connected to a power source and an SWR meter of good accuracy, the dummy load presents an SWR of unity (1-to-1) at 2 meters. At 430 MHz, the SWR meter reads about 1.1-to-1, indicating the load is slightly reactive at this frequency.

The dummy load is very useful in making SWR measurements on a new antenna when solid state VHF equipment is used. By using the load as an attenuator (J1 input, J3 output) the station equipment is protected against high SWR conditions in the antenna system during antenna tests. The attenuator also stabilizes and simplifies SWR measurements when placed between the SWR meter and the transmitter. For prolonged use as a dummy load, the cover should be removed from the box to improve cooling.

Care & Feeding of the "Ni-Cad" Battery

The nickel-cadmium (Ni-Cad) battery is a high efficiency cell capable of being recharged hundreds of times under the proper circumstances (Fig. 5). It is used in most of the popular, imported VHF HTs (hand-held transceivers). The nominal potential of the Ni-Cad is 1.25 volts, and a number of them are often connected in series to provide a higher operating voltage.

The common Ni-Cad is a non-vented, sealed cell which resembles a conventional carbon-zinc dry cell in appearance. The terminal voltage of a Ni-Cad cell varies with the state of charge and should run between 1.25 and 1.30 volts on open (no-load) circuit.

The amount of time a Ni-Cad will sustain its charge is a function of the discharge time, which is determined by the amount of current drawn. The cells are rated in **ampere-hours** (a-h), the product of current drain and time. The AA cells are available in three ratings, depending upon the manufacturer and cell style: 0.4, 0.5, or 0.7 ampere-hours. The C cell has a rating of 2 a-h; the D cell, 4 a-h; and the F cell, 7 a-h.

The ampere-hour rating indicates the amount of current drawn in amperes that will discharge the cell to one volt in one hour. In real life, the ampere-hour rating merely gives the user an idea of the cell capacity as compared to cells of another size.

The Ni-Cad battery has a finite storage life. If unused, the typical cell has a shelf-life of only a few months. The solution to this problem is to **trickle charge** the cell continually while it is stored. The charging current during long-time storage runs about 1/50 the a-h figure. Thus a AA cell

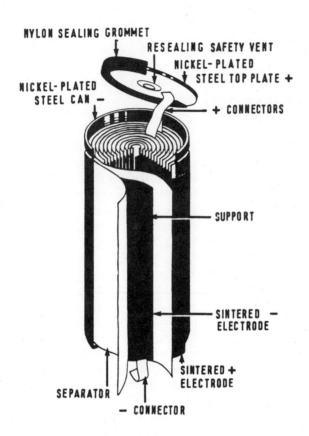

Fig.5 *Construction of a typical Ni-Cad cell. The nickel-cadmium cell is a sealed device. Oxygen produced during operation is recycled so there is no loss of electrolyte. Most cells have a safety vent than enables cell to release gas under heavy load and then to reseal automatically. Chemical action of the cell causes a temperature rise and it is necessary to limit charging current to prevent overheating and overcharge.*

with an a-h rating of 0.7 would be continually trickle charged when not in use at $0.7 \times 1/50 =$ about 15 milliamperes.

Charging the Ni-Cad Battery

When the Ni-Cad is discharged, a safe charging current is about 1/10 the a-h figure. The battery must be charged to 140 percent of capacity, so a charging time of 14 hours is required. For all Ni-Cad batteries the general charging rule is: charge at 1/10 the a-h rating for 14 hours.

The Ni-Cad battery has a charging "memory" in that it will not allow a deep discharge after repeated shallow discharges. For example, if a battery is repeatedly discharged to only 50 percent of its capacity, it will "remember" the 50 percent level as the fully discharged point. This makes the battery appear to have a premature failure. The battery may be restored to full capacity by recharging, then immediately discharging it. After a second charge, the memory pattern will be cleared.

Many amateurs keep two battery packs for their HTs. One is in the HT and the other is on trickle charge. When the pack in use runs low, it is swapped for the one on charge.

The Ni-Cad battery should never be left in a discharged condition as the internal plates deteriorate very rapidly. Keep the battery charged; trickle charge it when not in use and discharge it fully in use. The Ni-Cad will then have a long and useful life.

How to Protect Your Mobile FM Equipment

Some important points concerning the mobile use of your new VHF equipment should be mentioned:

1- It is possible to damage the FM transceiver if the equipment is left turned on when the vehicle engine is started. Large variations in the voltage of the car electric system may be created, and voltage "spikes" generated by the starting process could damage your FM gear.

2- Make sure your transceiver is not connected to the vehicle electrical system with reverse polarity. This can raise havoc with equipment components. Some (but not all) transceivers have internal protection circuits which block the flow of current if the user mistakenly cross-connects the power leads. But don't bet on it!

3- Don't inadvertently slam the door on the coax cable to the antenna! Dumb? Sure it is, but it happens more than anyone admits—even to old timers! One door slam on a cable could short it out, or break the conductors. If your cable must go through a door or trunk opening, make sure it is protected by the flexible rubber seal.

4- Make sure cables and control wires do not interfere with the braking and steering of the vehicle. Keep all equipment and wires out of the critical area surrounding the driver, brake pedal and accelerator.

5- If you have to drill holes in your vehicle to mount VHF gear, or to route cables, it is smart to see what is on the other side of the proposed hole! Auto manufacturers place cables, fuel, and hydraulic lines in odd places and a slip of the drill can create havoc before it can be stopped.

6- Remember most VHF equipment draws appreciable current from the automotive electrical system. Make your power leads heavy and short. A 40 watt transceiver, for example, may draw up to 8 amperes. It is OK to use the cigarette lighter plug for a 10 watt transceiver, but the larger unit requires that separate, heavy leads (no. 10, or better) run directly from the unit to the vehicle battery.

7- Take care of your car battery. Terminals and mounting frame should be kept clean of corrosion. The electrolyte level should be monitored if the battery is not sealed. The specific gravity should be checked with a hydrometer. A reading of 1.275 indicates a full charge, whereas a reading at or below 1.150 indicates the battery is close to the discharge point. Make sure the vehicle's voltage regulator is functioning properly.

8- Beware of theft! Stolen ham gear is a severe and growing problem. Lock your car! Remove the transceiver and take it with you, or lock it in the trunk when your vehicle is unattended. A mobile antenna is a sure way to attract thieves. Remove it if you can, and place it in the trunk. If you cannot remove your transceiver, place a dark cloth over it so that it is less conspicuous to a passerby.

9- Don't forget that call letter license plates and other radio-type identification draw the attention of thieves to your vehicle, so think twice before you identify your vehicle as being owned by a radio amateur. Mark all items of mobile equipment with an electric pencil so that they may be readily identified. Log the serial numbers of your equipment and keep the information in a safe place. If your equipment is stolen, this information will be of value to the police.

The "Mobile" Rig in the Home Station

For reasons unknown, many manufacturers of compact VHF transceivers term them "mobile transceivers" in their ads, possibly because they are designed to operate from a 12.6 volt automotive electrical system. Some newcomers to the ham game don't realize these "mobile" rigs are ideally suited for home station use if they are powered by a regulated 12.6 volt dc supply, running from the 120 volt, 60 Hz home electrical system. Regulated supplies are readily available at a nominal price and a 10 or 20 watt "mobile" transceiver using such a supply makes an ideal home station for the VHF novice or beginner.

Use Your CB Antenna on 10 Meters!

Yes, you can modify your CB antenna for 10 meter operation. It is necessary to scale down the original dimensions so the antenna resonates in the 10 meter band. Most CB antennas are cut for 27.1 MHz. It is necessary to retune them to about 28.3 MHz for operation in the lower half of the 10 meter band. The reduction ratio is the ratio of the frequencies, that is, 27.1 / 28.3, or about 0.96.

In the case of a quarter-wave ground plane antenna the reduction ratio indicates the vertical section and radials should all be shortened about 4-1/2 inches at the tips. A dipole antenna should be shortened 4-1/2 inches on leg, for a total of 9 inches, overall.

A Yagi beam requires that each element be reduced 4-1/2 inches on each tip. Element spacing need not be changed.

It is not so easy to reduce the elements in a Quad antenna. The overall length of wire in each loop should be shortened 18 inches. This will entail moving the tie points on each arm of the Quad to accomodate the smaller loop.

In the case of the 5/8-wave vertical antenna with radials, the radials are reduced 4-1/2 inches in length and the vertical section is reduced 11 inches.

If the antenna has a matching network, it will have to be readjusted for the lowest SWR at 28.3 MHz for the revised antenna. This can be accomplished with the aid of an SWR meter in the transmission line to the antenna.

Additional Reading

DeSoto, C.B.: "Two Hundred Meters and Down" (ARRL). A history of amateur radio covering the advance into the higher frequencies.

Staff: "The ARRL Operating Manual" (ARRL). Correct operating procedures and techniques.

Staff: "The ARRL Repeater Directory" (ARRL). Invaluable list of repeaters, frequencies and locations.

Davidoff, M.: "The Satellite Experimenter's Handbook" (ARRL). All about satellites from A to Z.

Nelson, W.R.: "Interference Handbook", Radio Publications Inc., 925 Sherwood Dr., Box 247, Lake Bluff, IL 60044. Covers all aspects of TVI, stereo, telephone and related forms of interference.

Orr, William: "The Radio Handbook", 23d edition. Howard W. Sams Co., 4300 W. 62d. St., Indianapolis, IN 46268. General theory, application and equipment design. Many VHF amplifiers covered in detail.

Orr, William: "All About HF Radio", Radio Publications Inc. Box 247, Lake Bluff, IL 60044. HF amateur radio from 160 meters to 10 meters.

Orr, William and Cowan, Stuart: "Beam Antenna Handbook", Radio Publications, Box 247, Lake Bluff, IL 60044. Yagi antenna theory, design and construction.

Lawson, J: "Yagi Antenna Design" (ARRL). A technical treatise on Yagi antennas.

Long and Keating: "The World of Satellite Television", Quantum Publishing, Inc., Box 310, Mendocino, CA 95460.

Kearman, J: "FM and Repeaters" (ARRL). An in-depth coverage of repeater operation.

Ingram, D: "RTTY Today", Universal Electronics, 4555 Groves Rd., Suite 3, Columbus, OH 43232. A detailed coverage of RTTY reception and transmission.

Publications listed (ARRL) may be obtained from American Radio Relay League, 225 Main St., Newington, CT 06111.

Acknowledgements
The author gratefully acknowledges the advice and support of the following in the preparation of this handbook:
Reid Brandon, W6MTF
Stuart Cowan, W2LX
Marty Davidoff, K2UBC
Dick Daniels, W4PUJ
Hal Jones, W6ZVV
Bob Locher, W9KNI
Herb Nelson Jr., W9IGL
Jim Stewart, WA4MVI
Bob Sutherland, W6PO
Special thanks to Jim Stewart and Howard W. Sams & Co., Indianapolis, IN for permission to reproduce the following illustrations from "VHF Radio Propagation", copyright 1978: Chap.1-Figs.14,20,22,23,25,26 and Table 1. Chap.2-Fig.1, Table 1. Chap.3-Fig. 8. Chap.4-Figs. 6,7.

INDEX

INDEX

OTHER BOOKS FOR RADIO AMATEURS, CB OPERATORS, SHORTWAVE LISTENERS, STUDENTS, & EXPERIMENTERS

ALL ABOUT CUBICAL QUAD ANTENNAS, by William I. Orr W6SAI and Stuart D. Cowan W2LX; 112 pages, 75 illustrations.

This well-known classic has been updated to include: new Quad designs; new dimension charts for every type of Quad from 6 to 80 meters; additional gain figures; an analysis of Quad vs. Yagi; Mini and Monster Quad designs; Delta, Swiss, and Birdcage Quads; and an improved Tri-Gamma match to feed a triband Quad efficiently with one transmission line. Also covered are feed systems and tuning procedures for maximum gain and minimum SWR. Much of this data has never before been published.

BEAM ANTENNA HANDBOOK, by William I. Orr W6SAI and Stuart D. Cowan W2LX; 271 pages, 205 illustrations.

This popular new edition gives you: correct dimensions for 6, 10, 15, 20, and 40 meter beams; data on triband and compact beams; the truth about beam height; SWR curves for popular beams from 6 to 40 meters; and comparisons of T-match, Gamma match, and direct feed. Describes tests to confirm if your beam is working properly, tells how to save money by building your own beam and balun, and discusses test instruments and how to use them. A "must" for the serious DX'er!

BETTER SHORTWAVE RECEPTION, by William I. Orr W6SAI and Stuart D. Cowan W2LX; 160 pages, 117 illustrations.

Tells what to look for when buying a shortwave receiver, 10 things to check when purchasing a used receiver, and explains how your radio receiver works. Teaches how to tune and adjust it for best performance. Covers receiver alignment to optimize reception. Discusses wave propagation, skip distance, and sunspot cycles. Describes scanning receivers, long distance TV and FM reception, and nine efficient receiving antennas. Tells how to hear foreign broadcasts, police, fire, aircraft, marine, weather, amateurs, CB'ers, and private business radio.

CARE AND FEEDING OF POWER GRID TUBES, by Robert Sutherland W6PO and Laboratory Staff of EIMAC; 166 pages, 83 illustrations.

This advanced, yet concise, handbook analyzes the operation of power grid tubes from audio to VHF. Design and application data for long tube life, maximum circuit stability, and peak efficiency are comprehensively covered. Includes thorough coverage of constant current curves and their application to circuit design. This is an ideal book for the serious communicator, one used as a text by leading engineering schools.

THE RADIO AMATEUR ANTENNA HANDBOOK, by William I. Orr W6SAI and Stuart D. Cowan W2LX; 191 pages, 147 illustrations.

This clearly written, easy to understand handbook contains a wealth of information about amateur antennas, from beams to baluns, tuners, and towers. The exclusive "Truth Table" gives you the actual dB gain of 10 popular antenna types. Describes how to build multiband vertical and horizontal antennas, Quads, Delta Quads, Mini-Quads, a Monster Quad, DX "slopers", triband beams, and VHF Quagi and log periodic Yagi beams. Dimensions are given for all antennas in English and Metric units. Tells how antenna height and location affect results and describes efficient antennas for areas with poor ground conductivity. Covers radials, coaxial cable loss, "bargain" coax, baluns, SWR meters, wind loading, tower hazards, and the advantages and disadvantages of crank-up tilt-over towers.

SIMPLE, LOW-COST WIRE ANTENNAS FOR RADIO AMATEURS, by William I. Orr W6SAI and Stuart D. Cowan, W2LX; 192 pages, 100 illustrations.

Now-another great handbook joins the famous Cubical Quad and Beam Antenna Handbooks. Provides complete instructions for building tested wire antennas from 2 through 160 meters-horizontal, vertical, multiband traps, and beam antennas. Describes a 3-band Novice dipole with only one feedline; the "folded Marconi" antenna for 40, 80, or 160 meters; "invisible" antennas for difficult locations-hidden, disguised, and disappearing antennas (the Dick Tracy Special, the CIA Special). Covers antenna tuners and baluns, and gives clear explanations of radiation resistance, impedance, radials, ground systems, and lightning protection. This is a truly practical handbook.

THE TRUTH ABOUT CB ANTENNAS, by William I. Orr W6SAI and Stuart D. Cowan W2LX; 240 pages, 145 illustrations.

Contains everything the CB'er needs to know to buy or build, install, and adjust efficient CB antennas for strong, reliable signals. A unique "Truth Table" shows the dB gain from 10 of the most popular CB antennas. The antenna is the key to clear, reliable communication but most CB antennas do not work near peak efficiency. Now, for the first time, this handbook gives clear informative instructions on antenna adjustment, exposes false claims about inferior antennas, and helps you make your antenna work. With exclusive and complete coverage of the "Monster Quad" beam, the "King" of CB antennas.

INTERFERENCE HANDBOOK, by William R. Nelson WA6FQG; Editor: William I. Orr W6SAI; 253 pages, 152 illustrations.

This timely book covers every radio frequency interference (RFI) problem, with solutions based on years of practical experience. Covers amateur radio, CB radio, and power line problems with proven solutions. Contains case histories and lists valuable tips for stereo and TV owners to cure interference. Covers mobile, telephone, CATV, and computer problems as well.

ALL ABOUT VERTICAL ANTENNAS, by William I. Orr W6SAI and Stuart D. Cowan W2LX; 192 pages, 95 illustrations.

Properly designed, built, and installed vertical antennas do a fine job in small places. This clear, well illustrated book covers the design, construction, installation, and operation of 52 vertical antennas: efficient Marconi antennas for 80 and 160 meters, multiband verticals, vertical loops, phased arrays, and shunt-fed towers. Also described are "radio" and electrical grounds, matching systems, tuners, loading coils, and TVI, plus the precautions necessary to protect yourself, your home, and your equipment from lightning damage . . . and much more! It's the most practical, authoritative vertical handbook published.

These popular handbooks save you time, trouble, and money in getting the most out of your equipment and your hobby. They condense years of study and successful experience into clear and interesting texts to help you obtain maximum results.

Radio Publications handbooks are available at better electronic dealers and bookstores everywhere. See your nearest dealer or write for free catalog.

RADIO PUBLICATIONS INC.
925 SHERWOOD DRIVE, BOX 247
LAKE BLUFF, IL 60044, USA
(312) 295−0060